Introductory Statistics

Concepts, Models, and Applications

Second Edition

David Stockburger

Southwest Missouri State University
Springfield, Missouri

ATOMICdogPUBLISHING.COM

Introductory Statistics, 2e
David Stockburger

Executive Editors:
Michele Baird, Maureen Staudt, and Michael Stranz

Executive Marketing Manager:
Rob Bloom

Developmental Editor:
Sarah Blasco

Sr. Marketing Coordinators:
Lindsay Annett and Sara Mercurio

Production/Manufacturing Manager:
Donna M. Brown

Production Editorial Manager:
Dan Plofchan

Rights and Permissions Specialist:
Kalina Hintz

Cover Design:
Red Hanger Design

Compositor:
Putman Productions

© 2008, 2001 Thomson Custom Solutions, a part of the Thomson Corporation. Thomson, and the Star logo, are trademarks used herein under license.

Printed in the
United States of America
1 2 3 4 5 6 7 8 10 09 08 07

For more information, please contact Thomson Custom Solutions, 5191 Natorp Boulevard, Mason, OH 45040. Or you can visit our Internet site at http://www.thomsoncustom.com

ALL RIGHTS RESERVED. No part of this work covered by the copyright hereon may be reproduced or used in any form or by any means — graphic, electronic, or mechanical, including photocopying, recording, taping, Web distribution or information storage and retrieval systems — without the written permission of the publisher.

For permission to use material from this text or product, contact us by:
Tel (800) 730-2214
Fax (800) 730 2215
www.thomsonrights.com

The Adaptable Courseware Program consists of products and additions to existing Thomson products that are produced from camera-ready copy. Peer review, class testing, and accuracy are primarily the responsibility of the author(s).

Introductory Statistics/David Stockburger – 2nd Edition

Book:
ISBN-10: 1-931-44202-9
ISBN-13: 978-1-931-44202-2

Package:
ISBN-10: 1-931-44246-0
ISBN-13: 978-1-931-44246-6

International Divisions List

Asia (Including India):
Thomson Learning
(a division of Thomson Asia Pte Ltd)
5 Shenton Way #01-01
UIC Building
Singapore 068808
Tel: (65) 6410-1200
Fax: (65) 6410-1208

Australia/New Zealand:
Thomson Learning Australia
102 Dodds Street
Southbank, Victoria 3006
Australia

Latin America:
Thomson Learning
Seneca 53
Colonia Polano
11560 Mexico, D.F., Mexico
Tel (525) 281-2906
Fax (525) 281-2656

Canada:
Thomson Nelson
1120 Birchmount Road
Toronto, Ontario
Canada M1K 5G4
Tel (416) 752-9100
Fax (416) 752-8102

UK/Europe/Middle East/Africa:
Thomson Learning
High Holborn House
50-51 Bedford Row
London, WC1R 4L$
United Kingdom
Tel 44 (020) 7067-2500
Fax 44 (020) 7067-2600

Spain (Includes Portugal):
Thomson Paraninfo
Calle Magallanes 25
28015 Madrid
España
Tel 34 (0)91 446-3350
Fax 34 (0)91 445-621

Contents

Atomic Dog publishes unique products that give you the choice of both online and print versions of your text materials. We do this so that you have the flexibility to choose which combination of resources works best for you. For those who use the online and print versions together, we numbered the primary heads and subheads in each chapter of both versions the same. For example, the first primary head in Chapter 1 is labeled 1-1, the second primary head in this chapter is labeled 1-2, and so on. The subheads build from the designation of their corresponding primary head: 1-1a, 1-1b, etc.

This numbering system is designed to make moving between the online and print versions as seamless as possible. So if your instructor tells you to read the material in 2-3 and 2-4 for tomorrow's assignment, you'll know that the information appears in Chapter 2 of both the online and print versions of the text.

Preface

It has been said that some texts are written to impress one's colleagues and others are written for students. This one is written for students. It is neither a mathematical treatise nor a cookbook. Instead of complicated mathematical proofs, I have written a book about mathematical ideas. I have substituted examples for proofs and require that the reader "believe!" on more than one occasion. The result is a text that can be understood by students. A grasp of the fundamental ideas presented in this text will prepare the student for a much more thorough treatment of statistics in a later course.

I have titled my text *Introductory Statistics: Concepts, Models, and Applications* for a reason. The order of the words in the subtitle is critical. I believe that without a fundamental understanding of the "big picture," the student will get lost in the details of statistics. What I have tried to do with this text is to present a conceptual framework around the term "models." Rather than attempting to provide many applications and examples and then asking the student to deduce the concept, I introduce the concept and then provide a few examples of the application of the concept.

History of the Text

The publication of this text fulfills a vision that started more than 23 years ago. In that vision, I would write a statistics text that would be distributed and read on a computer. I started writing my text because I was not satisfied with the contents of the statistics texts at that time. My text was going to be different: It was going to explain the underlying concepts of statistics, explore controversial issues, and eliminate almost all computational formulas. I required my students to purchase statistical calculators and showed them how to use statistical packages on the university's mainframe computer to do much of the computation.

I began to realize my vision on a computer that I soldered together and that used a cassette tape player to store the document. I soon learned how to generate unique assignments for each student. I did so because I didn't see how a student could get much benefit from frantically copying his or her neighbor's homework assignment.

I wrote the text and had the university bookstore print it and distribute it to my students. As computers changed, so did the word processor, graphics generator, statistical package, statistical calculator, and printer used to generate the text. The text went through many iterations, each an improvement. I found that even slight changes like moving the chapter caused a ripple effect throughout the text. Creating a user interface for the normal curve table eliminated the need for the table and the text that described how to use the table. This also had a ripple effect in later chapters, with the result that it is possible to describe conceptually what is occurring without the need to bog the student down in mindless details about how to do it.

In the early 1990s, I distributed the text as a WordPerfect document using FTP (an early means of transferring files). It was too much trouble for anyone to download the document, access it using the right version of the word processor, and then print it. In addition, it was really hard to publicize its availability. When the World Wide Web (WWW) took off in the mid 1990s, I converted the material again, this time to HTML. I made the text freely available to anyone who was connected to the WWW and took steps to insure that its delivery system was adequate. I advertised the availability of the text on search engines. I explored the possibility of using simulations and interactive exercises to illustrate statistical principles. My thought was, "What good is a book if no one reads it?" It was my gift to the world.

For this edition, I have converted the text yet again, this time to XML. I want to thank my publisher, Alex von Rosenberg, for believing in the text and encouraging me to update the material and means of presentation. The text is no longer free, but additions have been made to enhance the value of the text. I was concerned about the viability of the text when I am no longer associated with the university. I no longer have such concerns.

Conceptual Framework of the Text
Computation

From the very beginning, this text was designed to take advantage of technological advances to compute statistics. In many ways, it is easier to write and teach computational procedures than it is to present the underlying logic of the method. I have selected what I consider to be the more demanding route for the student.

Assumptions

If statistics involves building models, and models are simplifications of the world, then building models necessarily involves simplifying assumptions. These assumptions may be explicit, as in the assumption that the residuals in a regression are normally distributed, or they may be implicit, as in the assumption that the numbers have meaning. It is critical that the assumptions made when constructing models be understood and be reasonable. On the other hand, it is wrong to become slaves to the assumptions, demanding that every assumption be fully satisfied, because they never will be.

In this text, I spend considerably more time than other texts attempting to make explicit some of the implicit assumptions underlying the use of numbers. In developing measurement theory, for example, I explore some of the nuances and issues surrounding meaning and numbers.

Probability Theory

The chapter on probability theory was one of the most difficult to write. The treatment given this topic in most introductory statistics texts is either bogged down in set theory and combinatorial mathematics or simply glossed over. I have attempted to integrate probability theory conceptually into the model-building schema presented in an earlier chapter. Rather than taking a divisive position as to the correct manner in which probabilities should be employed, I present the material in such a way that probabilities become a useful tool to help us make decisions about the world, no matter what the theoretical background.

Hypothesis Testing

In this text, I present classical hypothesis testing, not necessarily because I see it as the one true path to knowledge, but because it has proven to be a useful tool in assisting people to make decisions about which effects are real and which could be due to chance. It would be almost impossible to underestimate the influence classical hypothesis testing has had on the social sciences. I have real reservations, however, about becoming slaves to the almighty .05 level of significance and present a fairly lengthy discussion of the importance of considering the costs of various errors when doing hypothesis tests.

Who Should Take This Course

This course is designed for individuals who desire knowledge about some very important tools used by the behavioral sciences to understand the world. Some degree of mathematical sophistication is necessary to understand the material in this course. The prerequisite for this text is a first course in algebra, preferably at the college level. Students have succeeded with less, but it requires a greater than average amount of time and effort on their part. If there is any doubt about whether or not the reader can follow the mathematics, it is recommended that Chapter 3, "The Language of Algebra," be attempted. If the attempt is successful, then the material in the rest of the book will most likely also be mastered.

A Brief History of the Teaching of Statistics

Emphasis has been placed in several different directions during the past two decades with respect to the teaching of statistics in the behavioral sciences. The first, during the 1950s and perhaps early 1960s, saw a focus on computation. During this time, large electro-mechanical calculators were available in many statistics laboratories. These calculators usually had ten rows and ten columns of number keys, and addition, subtraction, multiplication, and division keys. If you were willing to pay enough money, you could get two accumulators on the top; one for sums and one for sum of squares. The calculators weighed between 50 and 100 pounds, filled a desktop, made a lot of noise, and cost more than $1,000. Needless to say, not many students carried one around in a backpack.

Because the focus was on computation, much effort was made by the writers of introductory textbooks on statistics to reduce the effort needed to perform statistical computations using these behemoths. This was the time period during which computational formulas were developed. These are formulas that simplify computation but give little insight into the meaning of the statistic. This is in contrast to definitional formulas that better describe the meaning of the statistic but are often a nightmare when doing large-scale computation. To make a long story short, students during this phase of the teaching of introductory statistics ended up knowing how to do the computations but had little insight into what they were doing, why they did it, or when to use it.

The next phase was a reaction to the first. Rather than computation, the emphasis was on meaning. This was also the time period of the "new math" in grade and high schools, when a strong mathematical treatment of the material was attempted. Unfortunately, many of the students in the behavioral sciences were unprepared for such an approach and ended up knowing neither the theory nor the computational procedure. Calculators available during this time were electronic, with some statistical functions available. They were still very expensive (more than $1,000) and would generally not fit in a briefcase. In most cases, the statistics texts still retained the old computational formulas.

The current trend is to attempt to make statistics as simple for the student as possible. An attitude of "I can make it easier, or more humorous, or flashier than anyone else has in the past" seems to exist among many introductory statistics textbooks. In some cases, this has resulted in the student sitting down for dinner and being served a hot fudge sundae. The goal is admirable, and in some cases achieved, but the fact remains that the material, and the logic underlying it, is difficult for most students.

Teaching Statistics
Using Technology

My philosophy is that the statistical calculator and statistical computer packages have eliminated the need for computational formulas, so those formulas have been eliminated from this text. Definitional formulas have been retained, and the student is asked to compute the statistic once or twice "by hand." Following that, all computation is done using the statistical features on the calculator.

This is analogous to the square root function on a calculator. How many times do people ever use the complex algorithm they learned in high school to find a square root? Seldom or never. It is my argument that the square root procedure should be eliminated from the mathematics classroom. It gives little or no insight into what a square root is or how it should be used. Since it takes only a few minutes to teach a student how to find a square root on a calculator, it is much better to spend the remaining classroom time discussing the meaning and the usefulness of the square root.

In addition, I've attempted to tie together the various aspects of statistics into a theoretical whole by closely examining the scientific method and its relationship to statistics. In particular, this is achieved by introducing the concept of models early in the course, and by demonstrating throughout the course how the various topics are all aspects of the same approach to knowing about the world.

Feelings about the Course

It is not unusual to hear students describe their past experience with a mathematics course something like: "I had an algebra class 10 years ago, I hated it, I got a 'D' in it, I put this course off until the last possible semester in my college career." With many students, statistics is initially greeted with mixed feelings of fear and anger—fear because of the reputation of mathematics courses, anger because the course is required for a major or minor and the student has had no choice in its selection. I believe that these emotions inhibit the learning of the material. It is my experience that before any real learning may take place, the student must relax, have some success with the material, and accept the course. It is the responsibility of the instructor to deal with these negative emotions. If this is not done, the course might just as well be taught by a combination of books and computers.

Another difficulty sometimes encountered by the instructor of a statistics course is the student who has done very well in almost all other courses and who has a desire to do very well in a statistics course. In some cases, it is the first time that course material does not come easily to the student, with the student not understanding everything the instructor says. Panic sets in, tears flow, or perhaps the student is simply never seen in a statistics classroom again. The student must be willing to accept the fact that a complete understanding of statistics may not be obtained in a single introductory course. Additional study, perhaps graduate work, is necessary to more fully grasp the area of statistics. This does not mean, however, the student has achieved nothing of value in the course. Statistics may be understood at many different levels; it is the purpose of this text to introduce the material in a manner such that the student achieves the first level.

Statistical Calculators and Statistical Packages
Statistical Calculators

A two-variable statistical calculator is necessary to follow along with many of the computational procedures presented in this text. Generally, if a calculator has two buttons, labeled $\bar{x}$ and $\bar{y}$, it will suffice for the text. No particular brand or model of calculator is specified, and it is the duty of the student to translate the general instructions included in this text into specific sequences of key presses for a particular calculator. Thus, a manual or at least an instructional sheet is an absolute necessity. More powerful, graphically oriented calculators will also work for the text, but it has been my experience that unless the student knows and loves such a calculator, the additional complexity is a distraction to understanding the basic concepts presented in the text.

Statistical Packages

Unlike the selection of a statistical calculator, where any number of different brands and models will work, this text uses a single statistical package, SPSS, to demonstrate computational procedures and presentation of results. SPSS is a copyrighted computer program developed by SPSS, Inc. (http://www.spss.com) and is widely used and available at both academic and business institutions. It is most likely available at your institution.

The purchase of SPSS for use on home computers can be done a number of different ways. Some educational institutions have site licenses with SPSS, Inc. that permit students to obtain personal copies of the full version at a nominal cost. If an institution does not have such a license, then it is often possible to purchase a "Student" version at a greatly discounted cost. The student version is basically a stripped-down copy of the full version, but will suffice for this text. A "Graduate Student" version, adding statistical functions and capabilities to the student version, may also be available at a higher cost. In general, the additional capabilities of the graduate student or full versions do not add much additional complexity from the user's perspective and are preferred if available at reasonable cost.

If these different versions—student, graduate student, and full—are not confusing enough, SPSS has iterated through numerous numbered versions. This text was written using the most current version, 10.0, available. The changes from versions 8.0 and 9.0 to 10.0 have been fairly minor, and readers should be able to follow along if they do not have the most current version.

Notation for Statistical Packages

Throughout the text, it is often necessary to document the procedure to compute a statistic using the SPSS statistical package. For example, in the following screen, the user has clicked "Analyze," followed by "Regression," and then "Linear." This sequence of clicks will be denoted as Analyze/Regression/Linear in this text.

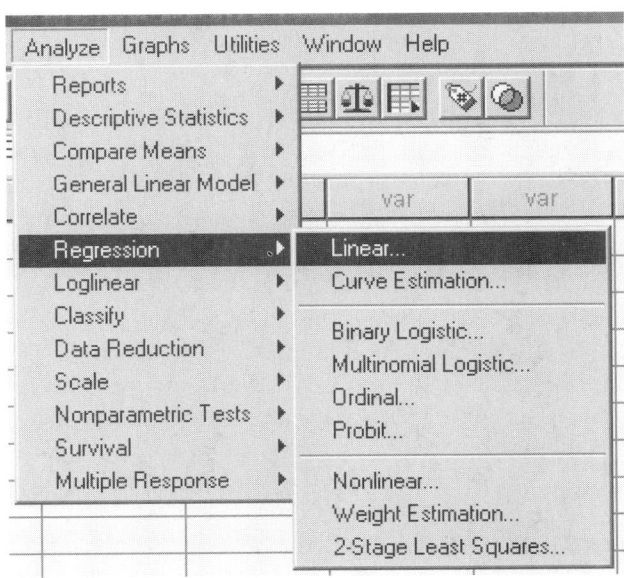

Objectives of the Text

Students successfully completing this course will

- Understand the relationship between statistics and the scientific method and how it applies to psychology and the behavioral sciences.
- Be able to read and understand the statistics presented in the professional literature.
- Be able to calculate and communicate statistical information to others.

Acknowledgments

I wrote this book for a number of reasons, the most important one being my students. Over a period of years, my approach to teaching introductory statistics began to deviate more and more from traditional textbooks. All too often, students would come up to me and say that they seemed to understand the material in class and thought they took good notes, but when they got home, the notes didn't seem to make much sense. Because the textbook I was using didn't seem to help much, I wrote this book. I took my lectures, added some documentation, and stirred everything with a word processor, with this book as the result.

This book is dedicated to all the students I have had over the years. Some made me think about the material in ways that I had not previously done, questioning the very basis of what this course was all about. Others were a different challenge in terms of how I could explain what I knew and understood in a manner in which they could comprehend. All have had an impact in one way or another.

Three students had a more direct input into the book and deserve special mention. Eve Shellenberger, an ex-English teacher, earned many quarters discovering various errors in earlier editions of the text. Craig Shealy took his editorial pencil to a very early draft and improved the quality greatly. Wendy Hoyt has corrected many errors in the online edition. To all, I am especially grateful.

I want to thank my former dean, Dr. Jim Layton, and my department head, Dr. Fred Maxwell, both of whom found the necessary funds for hardware and software acquisition to enable this project. Recently, I received funds from the Southwest Missouri State University academic vice president, Dr. Bruno Schmidt, to assist me in the transfer of this text from paper to online format.

About the Author

 David Stockburger is a professor of psychology at Southwest Missouri State University, where he has been teaching undergraduate and graduate statistics for 26 years. He earned his Ph.D. at Ohio State University in 1975 with a major area of mathematical psychology and a minor area of statistics. In his career, he has been involved in the application of technology to education, specifically statistics education, and has presented papers and written numerous articles on the topic. One of his proudest accomplishments is the faculty sponsorship of a student-run bulletin board system for six years before the web became a common appliance. When off campus, he resides with his wife in Springfield, Missouri, and enjoys golf, tennis, camping, traveling, and developing computer software in his leisure time.

Chapter

1

A Mayoral Fantasy

Key Terms

descriptive function of statistics

distorted sample

frequency distributions

inferential function of statistics

probable error

random sample

sample

sampling

statistics

Imagine, if you will, that you have just been elected mayor of a medium-sized city. You like your job; people recognize you on the street and show you the proper amount of respect; you are always being asked to out lunch and dinner, and so on. You want to keep your job as long as possible.

In addition to electing you mayor, the electorate voted for a new income tax at the last election. In an unprecedented show of support for your administration, the amount of the tax was left unspecified, to be decided by you (this is a fantasy!). You know the people of the city fairly well, however, and they would throw you out of office in a minute if you taxed them too much. If you set the tax rate too low, the effects might not be as immediate, as it takes some time for the city, police, and fire departments to deteriorate, but they are just as certain.

You have a pretty good idea of the amount of money needed to run the city. You do not, however, have more than a foggy notion of the distribution of income in your city. The IRS, being the IRS, refuses to cooperate. You decide to conduct a survey to find the necessary information.

1-1 A Full-Blown Fiasco

Since there are approximately 150,000 people in your city, you hire 150 students to conduct 1,000 surveys each. It takes significant time to hire and train the students to conduct the surveys. You decide to pay them $5.00 a survey, a considerable sum when the person being surveyed is a child with no income, but not much for the richest man in town who employs an army of CPAs. The bottom line is that it will cost approximately $750,000—close to three-quarters of a million dollars—to conduct this survey.

After a considerable period of time (because it takes time to conduct that many surveys), your secretary announces that the survey is complete. Boxes and boxes of forms are placed on your desk.

You begin your task of examining the figures. The first one is $33,967, the next is $13,048, the third is $309,339, and so on. Now the capacity for human short-term memory is approximately five to nine chunks (7 plus or minus 2). What this means is that by the time you are examining the tenth income, you have forgotten one of the previous incomes, unless you put the incomes in long-term memory. Placing 150,000 numbers in long-term memory is slightly overwhelming so you do not attempt that task.

1-2 Organizing and Describing the Data: An Alternative Ending

In an alternative ending to the fantasy, suppose you had at one time in your college days made it through the first half of an introductory statistics course. This part of the course covered the **descriptive function of statistics,** that is, procedures for organizing and describing sets of data.

Basically, there are two methods of describing data: pictures and numbers. Pictures of data are called **frequency distributions** and make the task of understanding sets of numbers cognitively palatable. Summary numbers may also be used to describe other numbers, and are called **statistics.** An understanding of what two or three of these summary numbers mean allows you to have a pretty good understanding of what the distribution of numbers looks like. In any case, it is easier to deal with two or three numbers than with 150,000. After organizing and describing the data, you make a decision about the amount of tax to implement. Everything seems to be going well until an investigative reporter from the local newspaper prints a story about the three-quarters of a million dollar cost of the survey. The irate citizens immediately start a recall petition. You resign the office in disgrace before you are thrown out.

1-3 Sampling: An Alternative Approach and a Happy Ending

If you had only completed the last part of the statistics course you would have understood the basic principles of the **inferential function of statistics.** Using inferential statistics, you can take a **sample** from the population, describe the numbers of the sample using descriptive statistics, and infer the population distribution. Granted, there is a risk of error involved, but if the risk can be minimized, the savings in time, effort, and money is well worth the risk.

In the mayoral fantasy, suppose that rather than surveying the entire population, you randomly selected 1,000 people to survey. This procedure is called **sampling** from the population and the individuals selected are called a sample. If each individual in the population is equally likely to be included in the sample, the sample is called a **random sample.**

Now, instead of 150 student surveyors, you only need to hire 10 surveyors, who each survey 100 citizens. The time taken to collect the data is a fraction of that taken to survey the entire population. Equally important, now the survey costs approximately $5,000, an amount that the taxpayers are more likely to accept.

At the completion of the data collection, the descriptive function of statistics is used to describe the 1,000 numbers, as it is still necessary to organize and describe the 1,000 numbers, but an additional analysis must be carried out to generalize (infer) from the sample to the whole population.

Some reflection on your part suggests that it is possible that the sample contained 1,000 of the richest individuals in your city. If this were the case, then the estimate of the amount of income to tax would be too high. Equally possible is the situation where 1,000 of the poorest individuals were included in the survey, in which case the estimate would be too low. These possibilities exist through no fault of yours or the procedure utilized. They are said to be due to chance; a **distorted sample** just happened to be selected.

The beauty of inferential statistics is that the amount of **probable error,** or likelihood of either of the these possibilities, may be specified. In this case, the possibility of either of these extreme situations actually occurring is so remote that they may be dismissed. However, the chance that there will be some error in our estimation procedure is pretty good. Inferential statistics will allow you to specify the amount of error with statements like, "I am 95 percent sure that the estimate is within $200 of the true value." You are willing to accept the risk of error and inexact information because the savings in time, effort, and money are so great.

At the conclusion of the fantasy a grateful citizenry makes you king (or queen). You receive a large salary increase and are elected to the position for life. You may continue this story any way that you like at this point....

Summary

This chapter presented the concept of statistics. You saw how statistics are useful in the real world, and took a brief look at the descriptive and inferential functions of statistics. You also were introduced to the concepts of samples and random samples.

Chapter

2

Models

Key Terms

formal language
measurement
model
natural language

parameters
parsimonious models
scientific method
symbolic model

syntax of language
variables

The knowledge and understanding that the scientist has about the world is often represented in the form of **models.** The **scientific method** is basically one of creating, verifying, and modifying models of the world. The goal of the scientific method is to simplify and explain the complexity and confusion of the world. The applied scientist and technologist then use the models of science to predict and control the world.

This book is about a particular set of models, called statistics, which social and behavioral scientists have found extremely useful. In fact, most of what social scientists know about the world rests on the foundations of statistical models. It is important, therefore, that social science students understand both the reasoning behind the models, and their application in the world.

2-1 Definition of a Model

A model is a representation containing the essential structure of some object or event in the real world.

The representation may take one of two major forms:

1. Physical, as in a model airplane or architect's model of a building

2. Symbolic, as in a natural language, a computer program, or a set of mathematical equations

In either form, certain characteristics are present by the nature of the definition of a model.

2-2 Characteristics of Models

There are two major characteristics of models:

1. Models are necessarily incomplete.

2. Models may be changed or manipulated with relative ease.

Because it is a representation, no model includes every aspect of the real world. If it did, it would no longer be a model. In order to create a model, a scientist must first make some assumptions about the essential structure and relationships of objects and/or events in the real world. These assumptions are about what is necessary or important to explain the phenomena.

For example, a behavioral scientist might want to model the time it takes a rat to run a maze. In creating the model the scientist might include such factors as how hungry the rat was, how often the rat had previously run the maze, and the activity level of the rat during the previous day. When constructing the model, the model-builder would also have to decide how these factors interacted. The scientist does not assume that only factors included in the model affect the behavior. Other factors might be the time of day, the experimenter who ran the rat, and the intelligence of the rat. The scientist might assume that these are not part of the "essential structure" of the time it takes a rat to run a maze. All the factors that are not included in the model will contribute to error in the predictions of the model.

To be useful, the model must be easier to manipulate than the real world is. The scientist or technician changes the model and observes the result, rather than doing a similar operation in the real world. He or she does this because it is simpler, more convenient, and/or the results might be catastrophic.

A race car designer, for example, might build a small model of a new design and test the model in a wind tunnel. Depending upon the results, the designer can then modify the model and retest the design. This process is much easier than building a complete car for every new design. The usefulness of this technique, however, depends on whether the essential structure of the wind resistance of the design was captured by the wind tunnel model.

Changing **symbolic models** is generally much easier than changing physical models, because all that is required is rewriting the model using different symbols. Determining the effects of such models is not always so easily accomplished. In fact, much of the discipline of mathematics is concerned with the effects of symbolic manipulation.

If the race car designer was able to capture the essential structure of the wind resistance of the design with a mathematical model or computer program, he or she would not have to build a physical model every time a new design was to be tested. All that would be required would be the substitution of different numbers or symbols into the mathematical model or computer program. As before, to be useful the model must capture the essential structure of the wind resistance.

The values that can be changed in a model to create different models are called **parameters.** In physical models, parameters are physical things. In the race car example, the designer might vary the length, degree of curvature, or weight distribution of the model. In symbolic models, parameters are represented by symbols. For example, in mathematical models parameters are most often represented by **variables.** Changes in the numbers assigned to the variables change the model.

2-3 The Language of Models

Of the two types of models, physical and symbolic, the latter is used much more often in science. Symbolic models are constructed using either a natural or formal language. Examples of **natural languages** include English, German, and Spanish. Examples of **formal languages** include mathematics, logic, and computer languages. Statistics as a model is constructed in a branch of the formal language of mathematics, algebra.

Natural and formal languages share a number of commonalties. First, they are both composed of a set of symbols, called the vocabulary of the language. English symbols take the form of words, such as those that appear on this page. Examples of algebraic symbols include 1, –3.42, X, +, =, and >.

The language consists of strings of symbols from the symbol set. Not all possible strings of symbols belong to the language. For instance, the following string of words is not recognized as a sentence, "Of is probability a model uncertainty," while the string of words "Probability is a model of uncertainty" is recognized almost immediately as being a sentence in the language. The set of rules to determine which strings of symbols form sentences and which do not is called the **syntax of the language.**

The syntax of natural languages is generally defined by common usage. That is, people who speak the language ordinarily agree on what is, and what is not, a sentence in the language. The rules of syntax are most often stated informally and imperfectly, for example, "noun phrase, verb, noun phrase."

The syntax of a formal language, on the other hand, may be stated with formal rules. Thus it is possible to determine whether or not a string of symbols forms a sentence in the language without resorting to users of the language. For example, the string "x + / y =" does not form a sentence in the language of algebra. It violates two rules in algebra: sentences cannot end in "=" and the symbols "+" and "/" cannot follow one another. The rules of syntax of algebra may be stated much more succinctly, as you will see in the next chapter.

Both natural and formal languages are characterized by the ability to transform a sentence in the language into a different sentence without changing the meaning or truth value of the string. For example, the active voice sentence "The dog chased the cat," may be transformed to the sentence "The cat was chased by the dog," by using the passive voice transformation. This transformation does not change the meaning of the sentence. In an analogous manner, the sentence "ax + ay" in algebra may be transformed to the sentence "a(x + y)" without changing the meaning of the sentence. Much of what has been taught as algebra consists of learning appropriate transformations, and the order in which to apply them.

The transformation process exists entirely within the realm of the language. The word proof will be reserved for this process. That is, it will be possible to prove that one algebraic sentence equals another. It will not be possible, however, to prove that a model is correct, because a model is never complete.

2-4 Model-Building in Science

The scientific method is a procedure for the construction and verification of models. After a problem is formulated, the process consists of four stages:

1. Simplification/Idealization.

As mentioned previously, a model contains the essential structure of objects or events. The first stage identifies the relevant features of the real world. The race car designer simplifies the design problem as one of car shape that reduces drag yet allows tires contact with the pavement. Steering, braking, and other mechanical systems are not part of the essential structure at this point.

2. Representation/Measurement.

The symbols in a formal language are given meaning as objects, events, or relationships in the real world. This is the process used in translating "word problems" to algebraic expressions in high school algebra. This process is called representation of the world. In statistics, the symbols of algebra (numbers) are given meaning in a process called **measurement.** In the case of the race car designer, the shape of the car might be represented using a piece of wood, metal, or plastic.

3. Manipulation/Transformation.

Sentences in the language are transformed into other statements in the language. In this manner implications of the model are derived. Using a physical model, the race car designer changes the shape of the piece of plastic or other material representing the car shape.

4. Verification.

Selected implications derived in the previous stage are compared with experiments or observations in the real world. Because of the idealization and simplification of the model-building process, no model can ever be in perfect agreement with the real world. In all cases, the important question is not whether the model is true, but whether the model is adequate for the purpose at hand. Model-building in science is a continuing process. New and more powerful models replace less powerful models, with "truth" being a closer approximation to the real world. Selecting the shape of the representation that worked best in the model, the race car designer builds a full-sized car. If the car wins, then the model is verified.

These four stages and their relationship to one another are illustrated here:

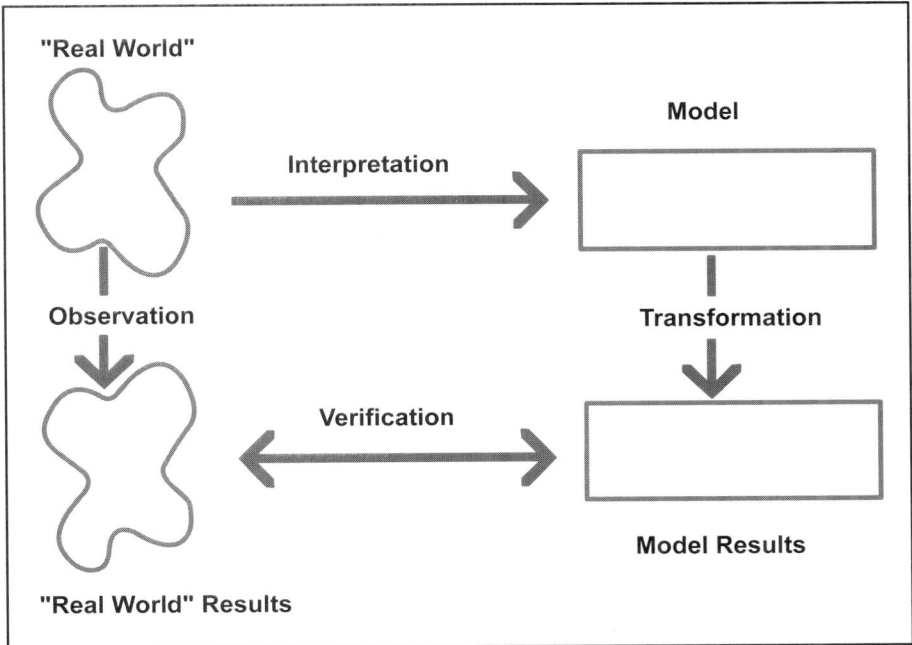

2-5 Power and Simplicity of Models

In general, the greater the number of simplifying assumptions made about the essential structure of the real world, the simpler the model. The goal of the scientist is to create simple models that have a great deal of explanatory power. Such models are called **parsimonious models.** In most cases, however, simple yet powerful models are not available to the social scientist. A trade-off occurs between the power of the model and the number of simplifying assumptions made about the world. A social or behavioral scientist must decide at what point the gain in the explanatory power of the model no longer warrants the additional complexity of the model.

2-6 Mathematical Models

The power of the mathematical model is derived from a number of sources. First, the language has been used extensively in the past and many models exist as examples. There are some very general models that can describe a large number of real world situations. In statistics, for example, the normal curve and the general linear model often serve the social scientist in many different situations. Second, many transformations are available in the language of mathematics.

Third, mathematics permits thoughts which are not easily expressed in other languages. For example, "What if I could travel approaching the speed of light?" or "What if I could flip this coin an infinite number of times?" In statistics these "what if" questions often take the form of questions like "What would happen if I took an infinite number of infinitely precise measurements?" or "What would happen if I repeated this experiment an infinite number of times?"

Finally, it is often possible to maximize or minimize the form of the model. Given that the essence of the real world has been captured by the model, what values of the parameters optimize (minimize or maximize) the model? For example, if the design of a race car can be accurately modeled using mathematics, what changes in design will result in the least possible wind resistance? Mathematical procedures are available that make these kinds of transformations possible.

2-7 Building a Better Boat: An Example of Model-Building

Suppose for a minute that you had lots of money, time, and sailing experience. Your goal in life is to build and race a 12-meter yacht that would win the America's Cup competition. How would you go about doing it?

Twelve-meter racing yachts do not have to be identical to compete in the America's Cup race, but there are certain restrictions on the length, weight, sail area, and other areas of boat design. Within these restrictions, there are variations that will change the handling and speed of the yacht through the water. The following two figures illustrate different approaches to keel design. The designer also has the option of installing a wing on the keel. If a wing is chosen, the decision of where it will be placed must be made.

Examples of boat keels.

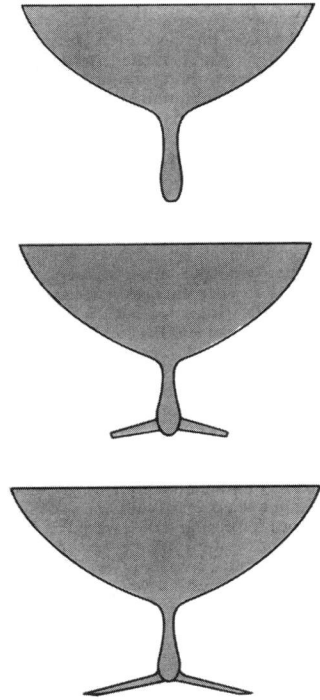

Different styles of "winged" keels.

You could hire a designer, have him or her draw up the plans, build the yacht, train a crew to sail it, and then compete in yachting races. All this would be fine, except it is a very time-consuming and expensive process. What happens if you don't have a very good boat? Do you start the whole process over again?

The scientific method suggests a different approach. If a physical model was constructed, and a string connected to weights was connected to the model through a pulley system, the time to drag the model from point A to point B could be measured. The hull shape could be changed using a knife and various weights. In this manner, many more different shapes could be attempted than if a whole new yacht had to be built to test every shape.

One of the problems with this physical model approach is that the designer never knows when to stop. That is, the designer never knows that if a slightly different shape were used, it might be faster than any of the shapes attempted up to that point. In any case the designer has to stop testing models and build the boat at some point in time.

Suppose the fastest hull shape was selected and the full-scale yacht was built. Suppose also that it didn't win. Does that make the model-building method wrong? Not necessarily. Perhaps the model did not represent enough of the essential structure of the real world to be useful. In examining the real world, it is noticed that racing yachts do not sail standing straight up in the water, but at some angle, depending upon the strength of the wind. In addition, the ocean has waves that necessarily change the dynamics of the movement of a hull through water.

If the physical model were pulled through a pool of wavy water at an angle, then the simulation would more closely mirror the real world. Assume that this is done, and the full-scale yacht built. It still doesn't win. What next?

One possible solution is the use of symbolic or mathematical models in the design of the hull and keel. The mathematical model uses parameters that allow the designer to change the shape of the simulated hull and keel by setting the values of the parameters to particular numbers. That is, a mathematical model of a hull and keel shape does not describe a particular shape, but a large number of possible shapes. When the parameters of the mathematical model of the hull shape are set to particular numbers, one of the possible hull shapes is specified. By sailing the simulated hull shape through simulated water, and measuring the simulated time it takes, the potential speed of a hull shape may be evaluated.

The advantage of creating a symbolic model over a physical model is that many more shapes may be assessed. By turning a computer on and letting it run all night, hundreds, thousands, or millions of shapes may be tested. It is sometimes possible to use mathematical techniques to find an *optimal model*, one that guarantees that within the modeling framework, no other hull shape will be faster. However, if the model does not include the possibility of a winged keel, the optimal model will never be discovered.

Suppose that these techniques are employed, and the yacht is built, but it still does not win. It may be that not enough of the real world was represented in the symbolic model. Perhaps the simulated hull must travel at an angle to the water and sail through waves. Capturing these conditions makes the model more complex, but is necessary if the model is going to be useful.

All the these modeling techniques were employed in building Team Dennis Conner's racing yacht *Stars and Stripes*. After initial computer simulation, a one-third scale model was constructed to work out the details of the design. The result of that model-building design process is history.

Summary

This chapter focused on modeling. A model is a representation containing the essential structure of some object or event in the real world. The representation may take two major forms: physical, as in a model car or an architect's model of a building, or symbolic, as in a natural language, a computer program, or a set of mathematical equations. There are two major characteristics of models: models are necessarily incomplete, and models may be changed or manipulated with relative ease.

Statistics use the model-building process to describe and make decisions about the world.

The scientific method is a procedure for the construction and verification of models. The process consists of four stages: simplification/idealization, representation/measurement, manipulation/transformation, and verification. The scientific method of model-building is a very powerful tool. Manipulation is the key to the process.

Chapter

3

The Language of Algebra

Key Terms

algebra
binomial expansion
delimiters

exponential notation
fractions
operators

transformational rules
variables

This section is intended as a review of the algebra necessary to understand the rest of this book, allowing you to gauge your mathematical sophistication relative to what is needed for the course. The individual without adequate mathematical training will need to spend more time with this chapter. The review of algebra is presented in a slightly different manner than has probably been experienced by most students, and may prove valuable even to the mathematically sophisticated reader.

Algebra is a formal symbolic language, composed of strings of symbols. Some strings of symbols form sentences within the language (X + Y = Z), while others do not (X + = Y Z). The set of rules that determines which strings belong to the language and which do not is called the syntax of the language. **Transformational rules** change a given sentence in the language into another sentence without changing the meaning of the sentence. This chapter first examines the symbol set of algebra, and then discusses syntax and transformational rules.

3-1 The Symbol Set of Algebra

The symbol set of algebra includes numbers, variables, operators, and delimiters. In combination, they define all possible sentences that may be created in the language.

3-1a Numbers

Numbers are analogous to proper nouns (such as names like Spot, Buttons, and Boots) in English. Some examples of numbers are

$$1, 2, 3, 4.89, -0.8965, -10090897.294, 0, \pi, e$$

Numbers may be either positive (+) or negative (-). If no sign is included, the number is positive. The two numbers at the end of the example list are called *universal constants*. The values for these constants are $\pi = 3.1416...$ and $e = 2.718....$

3-1b Variables

Variables are symbols that stand for any number. They are the common nouns within the language of algebra. Common nouns within the English language include words such as dog and cat that describes a general category of animals and not a particular dog or cat. Letters in the English alphabet most often represent variables, although Greek letters are sometimes used. Some example variables are

$$X, Y, Z, W, a, b, c, k, \sigma, \mu, r.$$

3-1c Operators

Other symbols, called **operators,** signify relationships between numbers and/or variables. Operators serve the same function as verbs in the English language. Some example operators in the language of algebra are

$$+, -, /, *, =, >, <, \geq, \leq.$$

Note that the * symbol is used instead of the "x" or "•" symbol for multiplication. This is common to many computer languages. The symbol $\geq$ is read as "greater than or equal to" and $\leq$ is read as "less than or equal to."

3-1d Delimiters

Delimiters are the punctuation marks in algebra. They let the reader know where one phrase or sentence ends and another begins. Example delimiters used in algebra are

$$(), [], \{ \}$$

In this course, only the "( )" symbols are used as delimiters, with the algebraic expressions being read from the innermost parentheses out.

Many statements in algebra can be constructed using only the symbols mentioned thus far, although other symbols exist. Some of these symbols will be discussed later in the book.

3-2 Syntax of the Language of Algebra

3-2a Creating Sentences

Sentences in algebra can be constructed using a few simple rules. The rules can be stated as replacement statements and are as follows:

Sentence	→	Phrase
Phrase	→	Number
Phrase	→	Variable
Phrase	→	Phrase Operator Phrase

The syntax of algebra.

Delimiters (parentheses) surround each phrase in order to keep the structure of the sentence straight. Sentences are constructed by creating a lower-order phrase and sequentially adding greater complexity, moving to higher-order levels. Here is an example of the construction of a complex algebraic sentence:

$$X + 3$$
$$7 * (X + 3)$$
$$(7 * (X + 3)) / (X * Y)$$
$$(P + Q) - ((7 * (X + 3)) / (X * Y))$$
$$((P + Q) - ((7 * (X + 3)) / (X * Y))) - 5.45$$

Statements such as this are rarely seen in algebra texts because rules exist to eliminate some of the parentheses and symbols in order to make reading the sentences easier. Generally these are *rules of precedence* where some operations (* and /) take precedence over others (+ and −).

3-2b Eliminating Parentheses

The following rules permit sentences written in the full form of algebra to be rewritten to make reading easier. Note that they are not always permitted when writing statements in computer languages such as PASCAL or BASIC.

1. The * symbol can be eliminated, along with the parentheses surrounding the phrase if the phrase does not include two numbers as subphrases. For example, (X * (Y− Z)) may be rewritten as X(Y − Z). However, 7 * 9 may *not* be rewritten as 79.

2. Any phrase connected with + or − may be rewritten without parentheses if the inside phrase is also connected with + or −. For example, ((X + Y) − 3) + Z may be rewritten as (X + Y) − 3 + Z. Continued application of this rule would result in the sentence X + Y − 3 + Z.

Sequential application of these rules may result in what appears to be a simpler sentence. For instance, take this sentence created from the earlier example:

$$((P + Q) - ((7 * (X + 3)) / (X * Y))) - 5.45$$

Apply Rule 1:

$$((P + Q) - 7(X + 3) / XY) - 5.45$$

Apply Rule 2:

$$P + Q - 7(X + 3) / XY - 5.45$$

Often these transformations are taken for granted and are already applied to algebraic sentences before they appear in algebra texts.

3-3 Transformations

Transformations are rules for rewriting sentences in the language of algebra without changing their meaning, or truth value. Much of what is taught in an algebra course consists of transformations. Transformations involve a range of operations from the simple addition of two numbers through the simplification of complex algebraic sentences. In all cases one algebraic sentence is replaced with another algebraic sentence that has the same meaning.

3-3a Numbers

When a phrase contains only numbers and an operator (such as 8 * 2), that phrase may be replaced by a single number (such as 16). These are the same rules that have been drilled into grade school students, at least up until the time of new math. The rule is to perform the operation and delete the parentheses. For example:

$$(8 + 2) = 10$$

$$(16 / 5) = 3.2$$

$$(3.875 - 2.624) = 1.251$$

The rules for dealing with negative numbers are sometimes imperfectly learned by students, so here's a quick review.

1. An even number of negative signs results in a positive number; an odd number of negative signs results in a negative number. For example:

$$-(-8) = -1 * -8 = 8 \text{ or } +8$$
$$-(-(-2)) = -1 * -1 * -2 = -2$$
$$-8 * - 9 * -2 = -144$$
$$-96 / -32 = 3$$

2. Adding a negative number is the same as subtracting a positive number. For example:

$$8 + (-2) = 8 - 2 = 6$$
$$-10 - (-7) = -10 + 7 = -3$$

3-3b Fractions

A second area that sometimes proves troublesome to students is that of fractions. **Fractions** are an algebraic phrase involving two numbers connected by the operator /; for example, 7/8. The top number or phrase is called the numerator, and the bottom number or phrase the denominator. One method of dealing with fractions that has gained considerable popularity since inexpensive calculators have become available is to do the division operation and then deal with decimal numbers. In what follows, two methods of dealing with fractions are illustrated. Select the method that is easiest for you.

Multiplication of fractions is relatively straightforward: multiply the numerators for the new numerator and the denominators for the new denominator. For example:

$$7/8 * 3/11 = (7 * 3) / (8 * 11) = 21/88 = 0.2386$$

Using decimals, the result would be

$$7/8 * 3/11 = .875 * .2727 = 0.2386$$

Division is similar to multiplication except the rule is to invert and multiply. For example:

$$(5/6) / (4/9) = (5/6) * (9/4) = (5 * 9) / (6 * 4) = 45/24 = 1.875$$

Using decimals, the result would be

$$.83333/.44444 = 1.875$$

Addition and subtraction with fractions first requires finding the least common denominator, adding (or subtracting) the numerators, and then placing the result over the least common denominator. For example:

$$(3/4) + (5/9) = (1 * (3/4)) + (1 * (5/9)) =$$
$$((9/9) * (3/4)) + ((4/4) * (5/9)) = 27/36 + 20/36 = 47/36 = 1.3056$$

Using the decimal form (simpler, in my opinion), the result is

$$3/4 + 5/9 = .75 + .5556 = 1.3056$$

Fractions have a special rewriting rule that sometimes allows an expression to be transformed to a simpler expression. If a similar phrase appears in both the numerator and the denominator of the fraction and these similar phrases are connected at the highest level by multiplication, then the similar phrases may be canceled. The rule is actually easier to demonstrate than to state, so here are some examples.

CORRECT application of the rule:

$$8X / 9X = 8/9$$
$$((X + 3) * (X - AY + Z)) / ((X + 3) * (Z - X)) = (X - AY + Z) / (Z - X)$$

INCORRECT application of the rule:

$$(X + Y) / X = Y$$

3-3c Exponential Notation

A number of rewriting rules exist within algebra to simplify an algebraic phrase with a shorthand notation of that phrase. **Exponential notation** is an example of a shorthand notational scheme. If a series of similar algebraic phrases are multiplied times one another, the expression may be rewritten with the phrase raised to a power. The power is the number of times the phrase is multiplied by itself and is written as a superscript of the phrase. For example:

$$8 * 8 * 8 * 8 * 8 * 8 = 8^6$$
$$(X - 4Y) * (X - 4Y) * (X - 4Y) * (X - 4Y) = (X - 4Y)^4$$

Some special rules apply to exponents. A negative exponent may be transformed to a positive exponent if the base is changed to one divided by the base. Here's a numerical example:

$$5^{-3} = (1/5)^3 = 0.008$$

A root of a number may be expressed as a base (the number) raised to the inverse of the root. For example:

$$\sqrt{16} = SQRT(16) = 16^{1/2}$$

When two phrases that have the same base are multiplied, the product is equal to the base raised to the sum of their exponents. The following examples illustrate this principle.

$$18^2 * 18^5 = 18^{5+2} = 18^7$$
$$(X + 3)^8 * (X + 3)^{-7} = (X + 3)^{8-7} = (X + 3)^1 = X + 3$$

It is possible to raise a decimal number to a decimal power, that is "funny" numbers may be raised to "funny" powers. For example:

$$3.44565^{1.234678} = 4.60635$$
$$245.967^{.7843} = 75.017867$$

Exponents such as the ones shown above may be evaluated using most scientific calculators. Generally these calculators work by first entering the base, such as 3.44565; clicking on a key, perhaps "XY"; entering the exponent, such as 1.234678; and then hitting another key, perhaps "=." Different calculators use different key combinations to achieve similar results, so consult the user's manual for the correct procedure for a given calculator. Even though most readers would not be able to simplify the above expression "by hand" because they never learned the algorithm to do this kind of simplification, that doesn't mean that it cannot be done. Somebody learned the algorithm, wrote a program to implement it, and made it available on calculators so that others might use it. Such expressions make perfect sense in the world of algebra and will be seen in later chapters.

3-3d Binomial Expansion

A special form of exponential notation, called **binomial expansion** occurs when a phrase connected with addition or subtraction operators is raised to the second power. Here are some examples of binomial expansion:

$$(X + Y)^2 = (X + Y) * (X + Y)$$
$$X^2 + XY + XY + Y^2 = X^2 + 2XY + Y^2$$
$$(X - Y)^2 = (X - Y) * (X - Y)$$
$$X^2 - XY - XY + Y^2 = X^2 - 2XY + Y^2$$

A more complex example of the preceding occurs when the phrase being squared has more than two terms.

$$(Y - a - bX)^2 = Y^2 + a^2 + (bX)^2 - 2aY - 2bXY + 2abX$$

3-3e Multiplication of Phrases

When two expressions are connected with the multiplication operator, it is often possible to "multiply through" and change the expression. In its simplest form, if a number or variable is multiplied by a phrase connected at the highest level with the addition or subtraction operator, the phrase may be rewritten as the variable or number times each term in the phrase. Again, it is easier to illustrate than to describe the rule:

$$a * (x + y + z) = ax + ay + az$$

If the number or variable is negative, the sign must be carried through all the resulting terms, as shown in the following example:

$$-y * (p + q - z) = -yp - yq - -yz$$
$$= -yp - yq + yz$$

Another example of the application of this rule occurs when the number -1 is not written in the expression, but rather inferred:

$$-(a + b - c) = (-1) * (a + b - c)$$
$$= -a - b - -c = -a - b + c$$

When two additive phrases are connected by multiplication, a more complex form of this rewriting rule may be applied. In this case, one phrase is multiplied by each term of the other phrase:

$$(c - d) * (x + y) = c * (x + y) - d * (x + y)$$
$$= cx + cy - dx - dy$$

Note that the binomial expansion discussed earlier was an application of this rewriting rule.

3-3f Factoring

A corollary to the rewriting rule for multiplication of phrases is factoring, or combining like terms. The rule may be stated as follows: If each term in a phrase connected at the highest level with the addition or subtraction operator contains a similar term, the similar term(s) may be factored out and multiplied times the remaining terms. It is the opposite of "multiplying through." Two examples follow:

$$ax + az - axy = a * (x + z - xy)$$
$$(a + z) * (p - q) - (a + z) * (x + y - 2z) = (a + z) * (p - q - x - y + 2z)$$

3-3g Sequential Application of Rewriting Rules to Simplify an Expression

Much of what is learned as algebra in high school and college consists of learning when to apply what rewriting rule to a sentence to simplify that sentence. Application of rewriting rules changes the form of the sentence, but not its meaning or truth value. Sometimes a sentence in algebra must be expanded before it may be simplified. Knowing when to apply a rewriting rule is often a matter of experience. As an exercise, simplify the following expression:

$$((X + Y)^2 + (X - Y)^2) / (X^2 + Y^2)$$

3-4 Evaluating Algebraic Sentences

A sentence in algebra is evaluated when the variables in the sentence are given specific values, or numbers. Two sentences are said to have similar truth values if they will always evaluate to equal values when evaluated for all possible numbers. For example, the sentence in the preceding exercise may be evaluated where $X = 4$ and $Y = 6$ to yield the following result:

$$((X + Y)^2 + (X - Y)^2) / (X^2 + Y^2)$$
$$((4 + 6)^2 + (4 - 6)^2) / (4^2 + 6^2)$$
$$(10^2 + -2^2) / (16 + 36)$$
$$(100 + 4) / (52)$$
$$104 / 52$$
$$2$$

The result should not surprise the student who correctly solved the preceding simplification. The result must *always* be equal to 2 as long as both X and Y are not zero. Note that the sentences are evaluated from the "innermost parentheses out," meaning that the evaluation of the sentence takes place in stages: phrases that are nested within the innermost or inside parentheses are evaluated before phrases that contain other phrases.

Summary

This chapter provided a review of the algebra necessary to understand the rest of this book. It focused on algebra as a formal symbolic language, composed of strings of symbols. The symbol set of algebra includes numbers, variables, operators, and delimiters, which, in combination, define all possible sentences that may be created in the language. You looked at the syntax of the language of algebra, and you reviewed transformations, the rules for rewriting algebraic sentences without changing their meaning. You also reviewed how to evaluate algebraic sentences.

Chapter

4

Measurement

Key Terms

intervals (property of)
interval scales
magnitude (property of)
measurement

measurement system
nominal scales
nominal-categorical scale
nominal-renaming scale

ordinal scales
ratio scales
rational zero (property of)
scale types

Measurement consists of rules for assigning numbers to attributes of objects based upon rules.

The language of algebra has no meaning in and of itself. The theoretical mathematician deals entirely within the realm of the formal language and is concerned with the structure and relationships within the language. The applied mathematician or statistician, on the other hand, is concerned not only with the language, but the relationship of the symbols in the language to real world objects and events. The concern about the meaning of mathematical symbols (numbers) is a concern about measurement.

By definition any set of rules for assigning numbers to attributes of objects is a **measurement system.** Not all measurement systems are equally useful in dealing with the world, however, and it is the function of the scientist to select those that are more useful. The physical and biological scientists generally have well-established, standardized systems of measurement. A scientist knows, for example, what is meant when a "ghundefelder fish" is described as 10.23 centimeters long and weighing 34.23 grams. The social scientist does not, as a general rule, have such established and recognized systems. A description of an individual as having 23 "units" of need for achievement does not evoke a great deal of recognition from most scientists. For this reason the social scientist, more than the physical or biological scientist, has been concerned about the nature and meaning of measurement systems.

4-1 Properties of Measurement Systems

S.S. Stevens (*Handbook of Experimental Psychology,* 1951) described properties of measurement systems that allowed decisions about the quality or goodness of a measurement technique. A property of a measurement system deals with the extent that the relationships which exist between the attributes of objects in the "real world" are preserved in the numbers that are assigned these objects. For example, if the attribute in question is height, then objects (people) in the "real world" have more or less of the attribute (height) than other objects (people). In a similar manner, numbers have relationships to other numbers. For example 59 is less than 62, 48 equals 48, and 73 is greater than 68. One property of a measurement system that measures height, then, is whether the relationships between heights in the "real world" are preserved in the numbers which are assigned to heights; that is, whether taller individuals are given bigger numbers.

Before describing in detail the properties of measurement systems, a means of symbolizing the preceding situation will be presented. You need not comprehend the following formalism to understand the issues involved in measurement, but mathematical formalism has a certain "beauty" which some students may appreciate.

Objects in the real world may be represented by O_i, where "O" is a shorthand notation for "object" and "i" is a subscript referring to which object is being described. The value of "i" may take on any integer number. For example O_1 is the first object, O_2 the second, O_3 the third, and so on. The symbol $M(O_i)$ will be used to symbolize the number, or measure (M), of any particular object that is assigned to that object by the system of rules—$M(O_1)$ being the number assigned to the first object, $M(O_2)$ the second, and so on. The expression $O_1 > O_2$ means that the first object has more of something in the "real world" than does the second. The expression $M(O_1) > M(O_2)$ means that the number assigned to the first object is greater than the number assigned to the second.

In mathematical terms, measurement is a functional mapping from the set of objects $\{O_i\}$ into the set of real numbers $\{M(O_i)\}$.

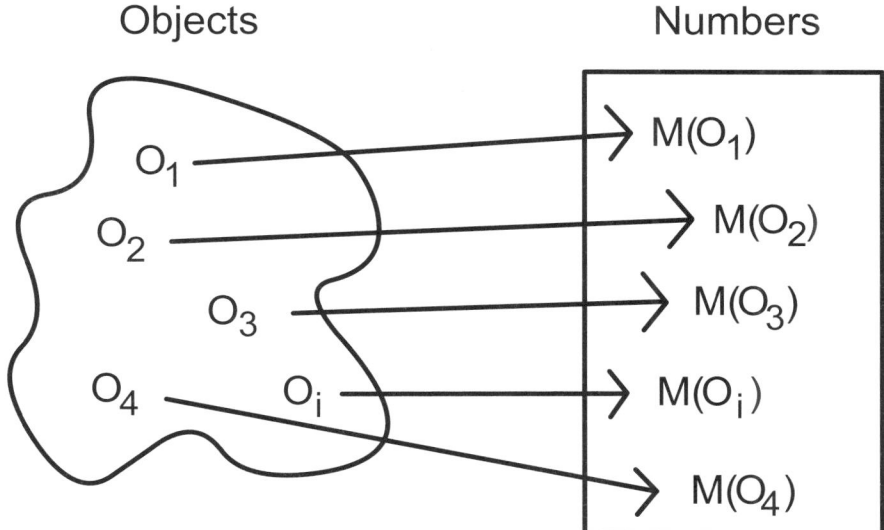

Objects   Numbers

A functional mapping of the set of objects into the set of numbers.

The goal of measurement systems is to structure the rules for assigning numbers to objects in such a way that the relationship between the objects is preserved in the numbers assigned to the objects. The different kinds of relationships preserved are called properties of the measurement system.

4-1a Magnitude

The **property of magnitude** exists when an object that has more of the attribute than another object is given a bigger number by the rule system. This relationship must hold for all objects in the "real world." Mathematically speaking, this property may be described as follows:

The property of magnitude *exists when for all i, j if $O_i > O_j$, then $M(O_i) > M(O_j)$.*

4-1b Intervals

The **property of intervals** is concerned with the relationship of differences between objects. If a measurement system possesses the property of intervals it means that the unit of measurement means the same thing throughout the scale of numbers. That is, an inch is an inch is an inch, no matter whether it falls immediately ahead or a mile down the road.

More precisely, an equal difference between two numbers reflects an equal difference in the "real world" between the objects that were assigned the numbers. In order to define the property of intervals in the mathematical notation, four objects are required: O_i, O_j, O_k, and O_l. The difference between objects is represented by the "–" sign; O_i–O_j refers to the actual "real world" difference between object i and object j, while $M(O_i)$–$M(O_j)$ refers to differences between numbers.

The property of intervals *exists, for all i, j, k, l*
if $O_i - O_j = O_k - O_l$ then $M(O_i) - M(O_j) = M(O_k) - M(O_l)$.

A corollary to the preceding definition is that if the number assigned to two pairs of objects are equally different, then the pairs of objects must be equally different in the real world. Mathematically it may be stated

If the property of intervals *exists if for all i, j, k, l*
if $M(O_i) - M(O_j) = M(O_k) - M(O_l)$, then $O_i - O_j = O_k - O_l$.

This provides the means to test whether a measurement system possesses the interval property, for if two pairs of objects are assigned numbers equally distant on the number scale, then it must be assumed that the objects are equally different in the real world. For example, in order for the first test in a statistics class to possess the interval property, it must be true that two students making scores of 23 and 28 respectively must reflect the same change in knowledge of statistics as two students making scores of 30 and 35.

The property of intervals is critical in terms of the ability to meaningfully use the mathematical operations "+" and "–". To the extent to which the property of intervals is not satisfied, any statistic that is produced by adding or subtracting numbers will be in error.

4-1c Rational Zero

A measurement system possesses the **property of rational zero** if an object that has none of the attribute in question is assigned the number zero by the system of rules. The object does not need to really exist in the "real world," as it is somewhat difficult to visualize a "man with no height." The requirement for a rational zero is this: if objects with none of the attribute did exist, would they be given the value zero? Defining O_0 as the object with none of the attribute in question, the definition of a rational zero becomes:

The property of rational zero exists if $M(O_0)=0$.

The property of rational zero is necessary for ratios between numbers to be meaningful. Only in a measurement system with a rational zero would it make sense to argue that a person with a score of 30 has twice as much of the attribute as a person with a score of 15. In many applications of statistics this property is not necessary to make meaningful inferences.

4-2 Scale Types

In the same article in which he proposed the properties of measurement systems, Stevens proposed four **scale types.** These scale types were *nominal, ordinal, interval,* and *ratio,* and each possessed different properties of measurement systems. Scale types were originally proposed as a way to classify measurement systems with respect to whether the properties would be preserved when various mathematical operations were used with the numbers that the system produced. For example, if a measurement system possessed the property of magnitude, would it still possess that property if all the numbers it produced were multiplied by three? Would it still posses that property if all the numbers it produced were scrambled? Others used scale types to classify measurement systems with respect to appropriateness for various kinds of statistical analysis. For example, they argued that unless a measurement system possessed the property of intervals, it was not appropriate to do many of the statistical analyses that will be discussed in later chapters.

Even though, in the opinion of the author, scale types have limited value as a conceptual framework for understanding measurement systems, they have an important historical value, and you may encounter these concepts and terms in the research literature.

4-2a Nominal Scales

Nominal scales are measurement systems that possess none of the three properties discussed earlier. Nominal scales may be further subdivided into two groups: renaming and categorical.

Nominal-renaming occurs when each object in the set is assigned a different number, that is, renamed with a number. Examples of nominal-renaming are Social Security numbers or numbers on the shirts of a baseball team. The former is necessary because different individuals have the same name, such as Mary Smith, and because computers have an easier time dealing with numbers rather than alphanumeric characters.

Nominal-categorical occurs when objects are grouped into subgroups and each object within a subgroup is given the same number. The subgroups must be mutually exclusive—that is, an object may not belong to more than one category or subgroup. An example of nominal-categorical measurement is grouping people into categories based upon stated political party preference (Republican, Democrat, or other) or upon gender (male or female). In the political party preference system Republicans might be assigned the number 1, Democrats 2, and Others 3, while in the latter females might be assigned the number 1 and males 2.

In general, it is meaningless to find means, standard deviations, correlation coefficients, and so forth when the data is nominal-categorical. If the mean for a sample based on the system of political party preferences was 1.89, you would not know whether most respondents were Democrats or whether Republicans and others were about equally split. This does not mean, however, that such systems of measurement are useless for, in combination with other measures, they can provide a great deal of information.

An exception to the rule of not finding statistics based on nominal-categorical scales types is when the data is dichotomous, or has two levels, such as Females = 1 and Males = 2. In this case, it is appropriate to both compute and interpret statistics that assume the interval property is met, because the single interval involved satisfies the requirement of the interval property.

4-2b Ordinal Scales

Ordinal scales are measurement systems that possess the property of magnitude, but not the property of intervals. The property of rational zero is not important if the property of intervals is not satisfied. Any time ordering, ranking, or rank ordering is involved, the possibility of an ordinal scale should be examined. As with a nominal scale, computation of most of the statistics described in the rest of the book is not appropriate when the scale type is ordinal. Rank-ordering people in a classroom according to height and assigning the shortest person the number 1, the next shortest person the number 2, and so on is an example of an ordinal scale.

4-2c Interval Scales

Interval scales are measurement systems that possess the properties of magnitude and intervals, but not the property of rational zero. It is appropriate to compute the statistics described in the rest of the book when the scale type is interval. Assigning people numbers by placing a book on their heads and observing where it crossed a tape measure placed on the wall could be considered an interval scale if the tape measure started out at a value other than zero. Assume that the dog chewed off the bottom two inches of the ruler, such that everyone was assigned a number two inches bigger than his or her actual height. In this case a person with no height would be assigned the number 2 and the measurement system would not possess the property of a rational zero.

4-2d Ratio Scales

Ratio scales are measurement systems that possess all three properties: magnitude, intervals, and rational zero. The added power of a rational zero allows ratios of numbers to be meaningfully interpreted, such as the ratio of John's height to Mary's height is 1.32, whereas this is not possible with interval scales. A system of measurement that assigned people a number for the attribute of height based on where a book crossed a tape measure placed on the wall could be considered a ratio scale if the tape measure started at zero. In this case, a person with no height would be assigned the number 0 and the rational zero property would be satisfied.

Note: Later in this chapter, it will be argued that these examples don't really work as the interval property of measurement systems is hardly ever completely satisfied.

It is at this point that most introductory statistics textbooks end the discussion of measurement systems and, in most cases, never talk about the topic again. Taking an opposite tack, some books organize the entire text around what is believed to be the appropriate statistical analysis for a particular scale type. The organization of measurement systems into a rigorous scale type classification leads to some considerable difficulties. The remaining portion of this chapter will point out those difficulties and a possible reconceptualization of measurement systems.

4-2e Exercises in Classifying Scale Types

The following is a list of different attributes and rules for assigning numbers to objects. Try to classify each of the different measurement systems into one of the four types of scales before reading any further.

- Your checking account number as a name for your account.

- Your checking account balance as a measure of the amount of money you have in that account.

- Your checking account balance as a measure of your wealth.

- The number you get from a machine (32, 33,...) as a measure of the time you arrived in line.

- The order in which you were eliminated in a spelling bee as a measure of your spelling ability.

- Your score on the first statistics test as a measure of your knowledge of statistics.

- Your score on an individual intelligence test as a measure of your intelligence.

- The distance around your forehead measured with a tape measure as a measure of your intelligence.

- A response to the statement "Abortion is a woman's right" where Strongly Disagree = 1, Disagree = 2, No Opinion = 3, Agree = 4, and Strongly Agree = 5, as a measure of attitude toward abortion.

4-3 The Myth of Scale Types

If you encountered difficulty in categorizing some of those descriptions, it does not mean that you lack understanding of the scale types. The problem resides in the method used to describe measurement systems; it simply does not work in many applied systems. J. Michell (1986, "Measurement Scales and Statistics: A Clash of Paradigms," *Psychological Bulletin,* 100–3, 398–407) presents a recent discussion of the controversy still present in psychology involving scale types and statistics.

Usually the difficulty begins in deciding the scale type of wealth as measured by bank account balance. Is it not possible that John has less money in the bank than Mary, but John has more wealth? Perhaps John has a pot of gold buried in his back yard, or perhaps he just bought a new car. Therefore the measurement system must be nominal-renaming. But if Mary has $1,000,000 in her account and John has –$10, isn't it more likely that Mary has more wealth? Doesn't knowing a person's bank account balance tell you *something* about her wealth? It just doesn't fit within the system.

Similar types of arguments may be presented with respect to the testing situations. Is it not possible that someone might score higher on a test yet know less or be less intelligent? Of course, maybe he didn't get enough sleep the night before or simply studied the wrong thing. On the other hand, maybe he was lucky—whenever he guessed, he got the correct answer. Should test scores not be used because they do not meet the requirements of an interval scale?

What about measuring intelligence with a tape measure? Many psychologists would argue that it is interval or ratio measurement, because it involves distance measured with a standard instrument. Not so. If your child were going to be placed in a special education classroom, would you prefer that the decision be based on the results of a standardized intelligence test or the distance around his or her forehead? The latter measure is nonsense, or almost entirely error.

4-3a The Ruler and Scale Types

Suppose a ruler was constructed in a nonindustrialized country in the following manner: peasants were given a stick of wood and sat painting interval markings and numbers on the wood. Would anything measured with this ruler be an interval scale? No, because no matter how good the peasants were at this task, the intervals would simply not be equal.

Suppose you purchase a ruler at a local department store. This ruler has been mass-produced at a factory in the United States. Would anything measured with this ruler be measured on an interval scale? No again, although it may be argued that it would certainly be closer than the painted ruler.

Finally, suppose you purchase a very precise Swiss caliper, measuring to the nearest 1/10,000 of an inch. Would it be possible to measure anything precisely on an interval scale using this instrument? Again the answer is no, although it is certainly the closest system presented thus far.

Suppose a molecular biochemist wanted to measure the size of a molecule. Would the Swiss caliper work? Is it not possible to think of situations where the painted ruler might work? Certainly the ruler made in the United States would be accurate enough to measure a room to decide how much carpet to order. The point is that in reality, unless a system is based on simple counting, an interval scale does not exist. The requirement that all measures be an interval or ratio scale before performing statistical operations makes no sense at all.

4-3b Toward a Reconceptualization of Scale Types

Measurement, as a process in which the symbols of the language of algebra are given meaning, is one aspect of the modeling process described in Chapter 2, "Models." Remembering the definition of a model as a representation of the "essential structure" of the real world, and not the complete structure, you would not expect that any system of measurement would be perfect. The argument that an interval scale does not exist in reality is not surprising viewed from this perspective.

The critical question is not whether a scale is nominal, ordinal, interval, or ratio, but rather whether it is useful for the purposes at hand. A measurement system may be "better" than another system because it more closely approximates the properties necessary for algebraic manipulation or costs less in terms of money or time, but no measurement system is perfect. In the view of the author, S.S. Stevens has greatly added to the understanding of measurement with the discussion of properties of measurement, but an unquestioning acceptance of scale types has blurred important concepts related to measurement for decades.

A discussion of error in measurement systems is perhaps a more fruitful manner of viewing measurement than scale types.

Different measurement systems exhibit greater or lesser amounts of different types of error. A complete discussion of measurement error remains for future study by the student.

The bottom line with respect to the theory of measurement that is of concern to the introductory statistics student is that certain assumptions are made, but never completely satisfied, with respect to the meaning of numbers. An understanding of these assumptions allows the student to better evaluate whether a particular system will be useful for the purposes intended.

Summary

The statistician is concerned not only with the language of algebra, but also with the relationship of the symbols in the language to real-world objects and events. The concern about the meaning of mathematical symbols (numbers) is a concern about measurement. Any set of rules for assigning numbers to attributes of objects is a measurement system, but not all measurement systems are equally useful in dealing with the world. While physical and biological scientists generally have well-established, standardized systems of measurement, the social scientist often does not. A description of an individual as having 45 "units" of need for security doesn't evoke much recognition from most scientists. Therefore, the social scientist, more than the physical or biological scientist, is concerned about the nature and meaning of measurement systems.

The goal of measurement systems is to structure the rules for assigning numbers to objects in such a way that the relationship between the objects is preserved in the numbers assigned to the objects. The different kinds of relationships preserved are called properties of the measurement system. This chapter discussed three such properties: magnitude, intervals, and rational zero.

The chapter also presented scale types, which were originally proposed as a way to classify measurement systems. Four scale types were discussed: nominal, ordinal, interval, and ratio, with nominal scales being subdivided into two groups: renaming and categorical. The critical question came to be not whether a scale was nominal, ordinal, interval, or ratio, but rather whether it was useful for the purposes at hand. A measurement system may be "better" than another system because it more closely approximates the properties necessary for algebraic manipulation or costs less in terms of money or time, but no measurement system is perfect.

A discussion of error in measurement systems was put forth as perhaps a more fruitful manner of viewing measurement than scale types. Different measurement systems exhibit greater or lesser amounts of different types of error. Basically, in the theory of measurement certain assumptions are made, but never completely satisfied, with respect to the meaning of numbers. An understanding of these assumptions allows you to better evaluate whether a particular measurement system will be useful for the purposes intended.

Frequency Distributions

Key Terms

absolute cumulative frequency
absolute frequency
absolute frequency polygon
bar graph
cumulative frequency
cumulative frequency polygon

data ink
frequency distribution
frequency polygon
frequency table
histogram
real limits

relative cumulative frequency
relative cumulative frequency
    polygon
relative frequency
relative frequency polygon

As discussed earlier, there are two major means of summarizing a set of numbers: pictures and summary numbers. Each method has advantages and disadvantages, and use of one method need not exclude the use of the other. This chapter describes drawing pictures of data, which are called **frequency distributions.**

5-1 Frequency Tables

The first step in drawing a frequency distribution is to construct a frequency table. A **frequency table** is a way of organizing the data by listing every possible score (including those not actually obtained in the sample) as a column of numbers and the frequency of occurrence of each score as another. Computing the frequency of a score is simply a matter of counting the number of times that score appears in the set of data. It is necessary to include scores with zero frequency in order to draw the **frequency polygons** correctly.

For example, consider the following set of 15 scores which were obtained by asking a class of students their shoe size, shoe width, and gender (male or female):

Data Table		
Shoe Size	Shoe Width	Gender
10.5	B	M
6.0	B	F
9.5	D	M
8.5	A	F
7.0	B	F
10.5	C	M
7.0	C	F
8.5	D	M
6.5	B	F
9.5	C	M
7.0	B	F
7.5	B	F
9.0	D	M
6.5	A	F
7.5	B	F

The following figure shows the same data entered into a data file in SPSS:

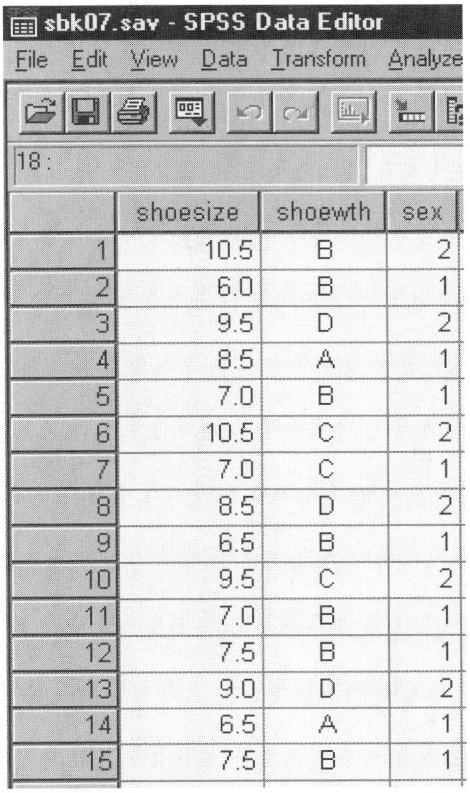

	shoesize	shoewth	sex
1	10.5	B	2
2	6.0	B	1
3	9.5	D	2
4	8.5	A	1
5	7.0	B	1
6	10.5	C	2
7	7.0	C	1
8	8.5	D	2
9	6.5	B	1
10	9.5	C	2
11	7.0	B	1
12	7.5	B	1
13	9.0	D	2
14	6.5	A	1
15	7.5	B	1

sbk07.sav - SPSS Data Editor
File Edit View Data Transform Analyze
18:

The first step in computing frequencies using SPSS.

To construct a frequency table, list all shoe sizes as a column of numbers, beginning with the smallest. The frequency of occurrence of each shoe size is written to the right, as the following frequency table of the example data shows.

Shoe Size	Abs. Freq.
6.0	1
6.5	2
7.0	3
7.5	2
8.0	0
8.5	2
9.0	1
9.5	2
10.0	0
10.5	2
Total	15

Note that the sum of the column of frequencies is equal to the number of scores or size of the sample (N = 15). This is a necessary, but not sufficient, property in order to ensure that the frequency table has been correctly calculated. It is not sufficient because two errors could have been made, canceling each other out.

While people think of their shoe size as a discrete unit, a shoe size is actually an interval of sizes. A given shoe size may be considered the midpoint of the interval. The **real limits** of the interval, the two points which function as cut-off points for a given shoe size, are the midpoints between the given shoe sizes. For example, a shoe size of 8.0 is really an interval of shoe sizes ranging from 7.75 to 8.25. The smaller value is called the lower real limit, while the larger is called the upper real limit. In each case, the limit is found by taking the midpoint between the nearest score values. For example, the lower limit of 7.75 was found as the average (midpoint) of 7.5 and 8.0 by adding the values together and dividing by two (7.5 + 8.0) / 2 = 15.5 / 2 = 7.75. A similar operation was performed to find the upper real limit of 8.25, that is, the midpoint of 8.0 and 8.5.

To generate a frequency table using the SPSS package, select Analyze/Descriptive Statistics/Frequencies, as illustrated in the following figure:

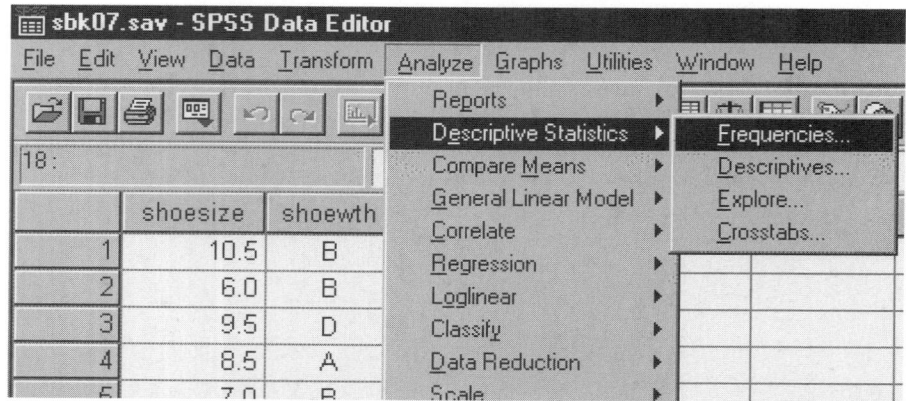

The second step in computing frequencies using SPSS.

In the frequencies box, select the variable name used for shoe size, and select the options shown in the following figure. Then click Continue and OK:

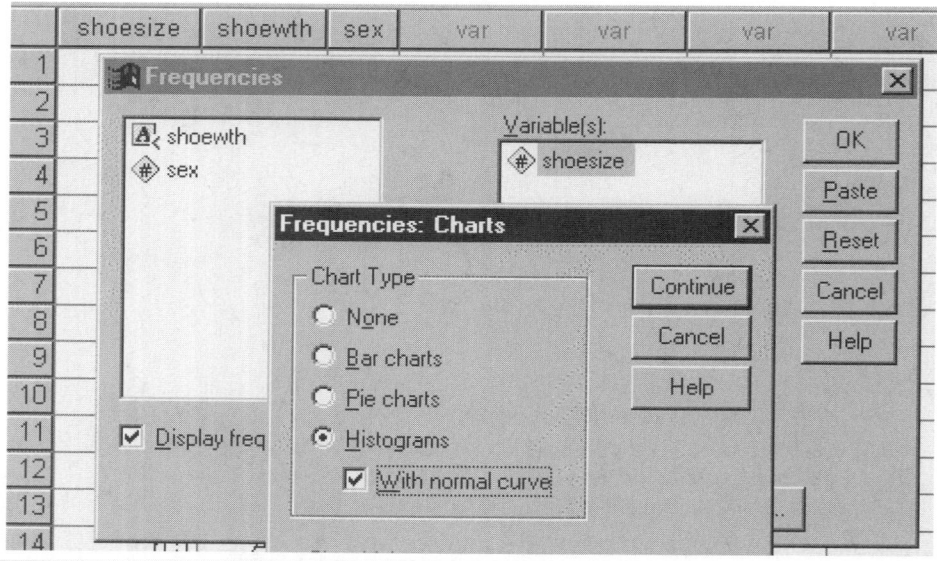

The third step in computing frequencies using SPSS.

The listing of the results of the analysis should contain the following:

Shoe Size		Frequency	Percent	Valid Percent	Cumulative Percent
Valid	6.0	1	6.7	6.7	6.7
	6.5	2	13.3	13.3	20.0
	7.0	3	20.0	20.0	40.0
	7.5	2	13.3	13.3	53.3
	8.5	2	13.3	13.3	66.7
	9.0	1	6.7	6.7	73.3
	9.5	2	13.3	13.3	86.7
	10.5	2	13.3	13.3	100.0
	Total	15	100.0	100.0	
Total		15	100.0		

Frequencies output in SPSS.

5-2 Frequency Distributions

The information contained in the frequency table may be transformed to a graphical or pictorial form. No information is gained or lost in this transformation, but the human information processing system often finds the graphical or pictorial presentation easier to comprehend. There are two major means of drawing a graph: histograms and frequency polygons. The choice of method is often a matter of convention, although there are times when one or the other is clearly the appropriate choice.

5-2a Histograms

A **histogram** is a graph drawn by plotting the scores (midpoints) on the X-axis and the frequencies on the Y-axis. A bar is drawn for each score value, the width of the bar corresponding to the real limits of the interval and the height corresponding to the frequency of the occurrence of the score value. The following histogram depicts the shoe size example data. Note that although there were no individuals in the example with shoe sizes of 8.0 or 10.0, those values are still included on the X-axis, with the bar for these values having no height.

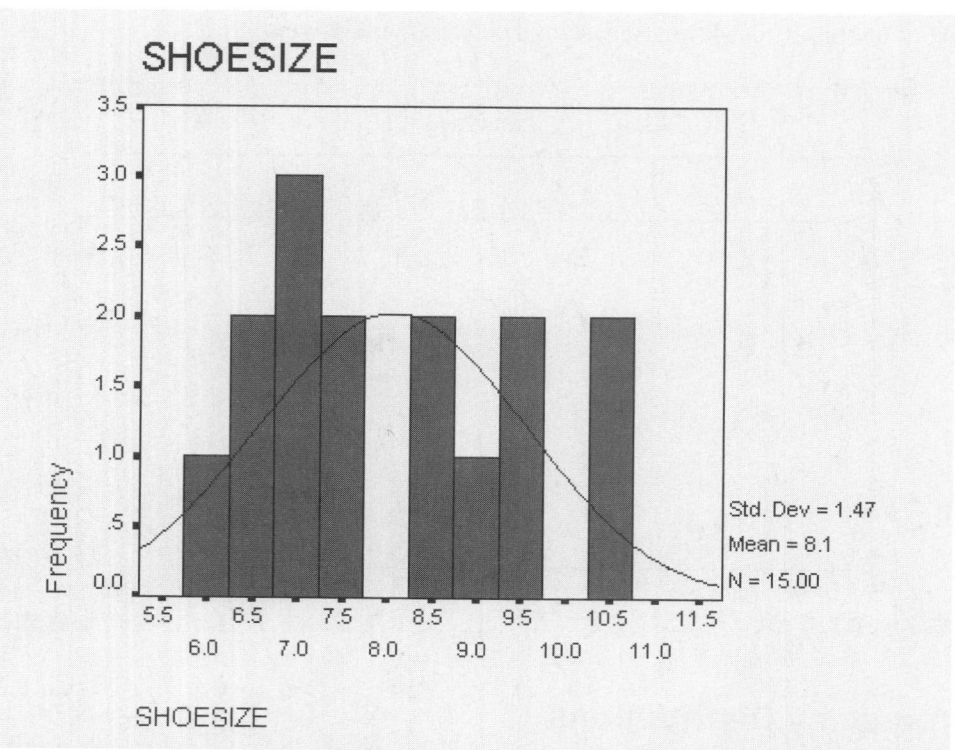

Example histogram from SPSS Frequencies command.

This histogram was drawn using the SPSS computer package. Included in the output from the Frequencies command was a histogram of shoe size. Unfortunately, the program must be edited in order to generate a figure like the one shown. To edit a figure in the listing file, place the cursor (arrow) on the figure and click the right mouse button. When a menu appears, select the last entry on the list, as the following figure shows:

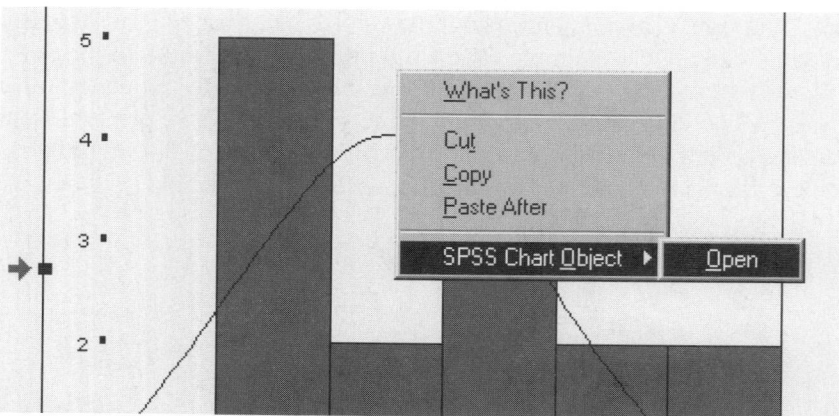

Modifying a chart in SPSS.

Edit the graph by selecting the options shown in the following figure:

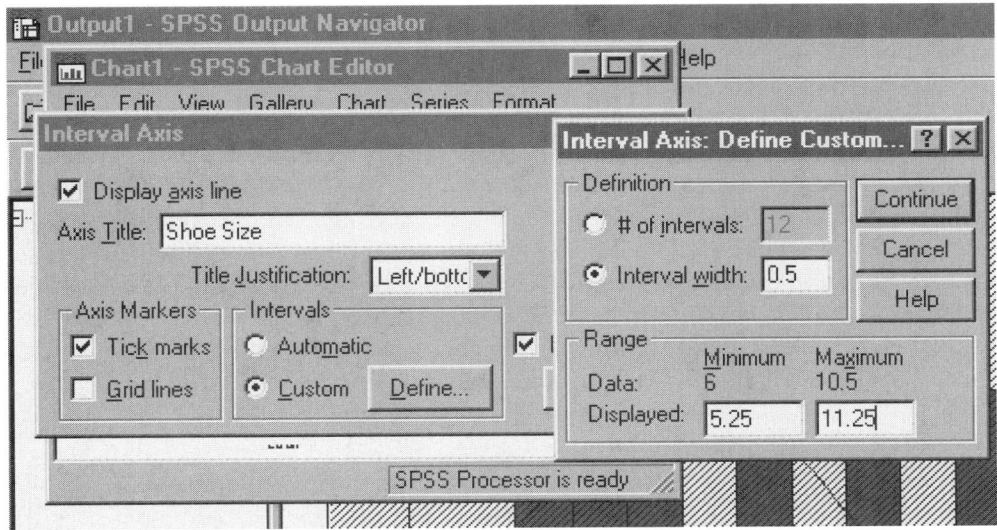

Changing the Interval Axis in the SPSS chart editor.

If the data are nominal-categorical in nature, the graph is similar, except that the bars do not touch, and the graph is called a **bar graph.** The following example presents the data for shoe width, assuming that it is not interval in nature. It was drawn using the example SPSS data file and the Bar Graph command.

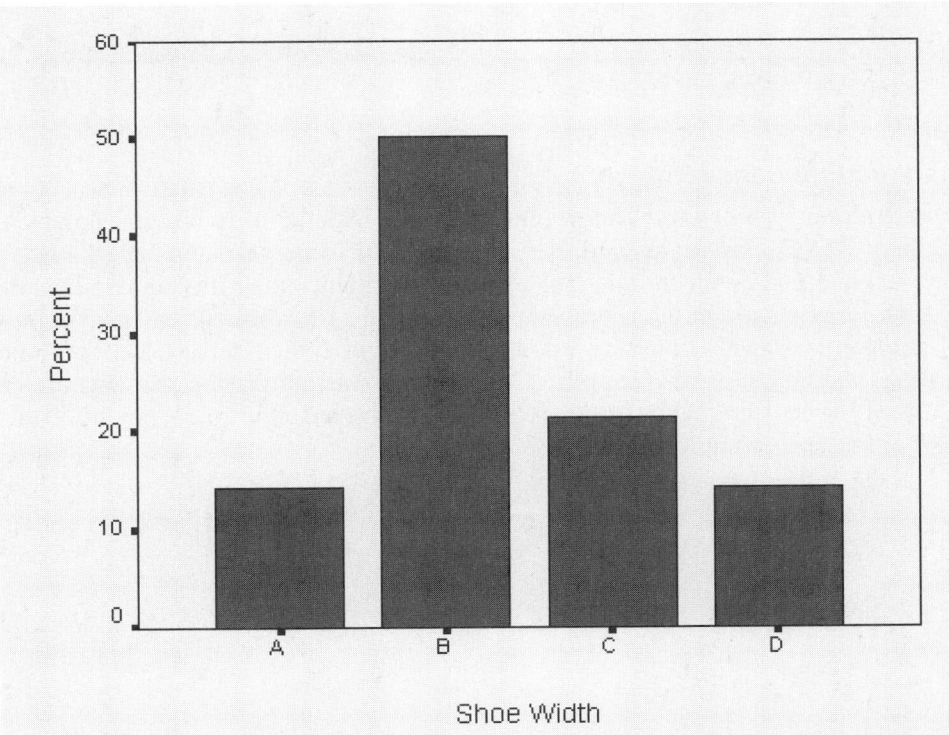

Histogram of shoe widths.

When the data are nominal-categorical in form, the bar graph is the only appropriate form for the picture of the data. When the data may be assumed to be interval, then the histogram can sometimes have a large number of lines, called **data ink,** which make the comprehension of the graph difficult. A frequency polygon is often preferred in those cases because much less ink is needed to present the same amount of information.

In some instances artists attempt to "enhance" a histogram by adding extraneous data ink. Two examples (see the following figure) of this sort of excess were taken from the local newspaper. In the first, the arm and building add no information to the illustration. San Francisco is practically hidden, and no building is presented for Honolulu. In the second, the later date is presented spatially before the earlier date and the size of the "bar" or window in this case has no relationship to the number being portrayed. These types of renderings should be avoided at all costs by anyone who in the slightest stretch of imagination might call himself "statistically sophisticated." An excellent source of information about presenting statistical data in picture form is Edward R. Tuft's *The Visual Display of Quantitative Information* (1983, Graphics Press).

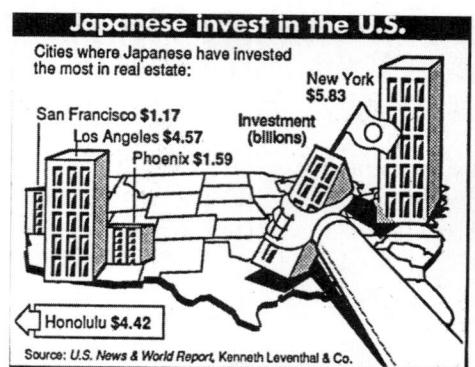 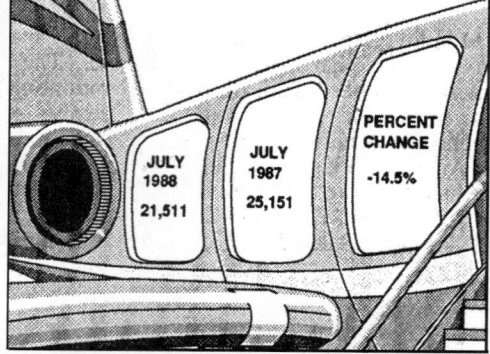

An example of a graph with a great deal of non-data ink.

5-2b Absolute Frequency Polygons

An **absolute frequency polygon** is drawn exactly like a histogram except that points are drawn rather than bars. The X-axis begins with the midpoint of the interval immediately lower than the lowest interval, and ends with the interval immediately higher than the highest interval. In the example, this would mean that the score values of 5.5 and 11.0 would appear on the X-axis. The frequency polygon is drawn by plotting a point on the graph at the intersection of the midpoint of the interval and the height of the frequency. When the points are plotted, the dots are connected with lines, resulting in a frequency polygon. The following absolute frequency polygon presents the data in the shoe size example.

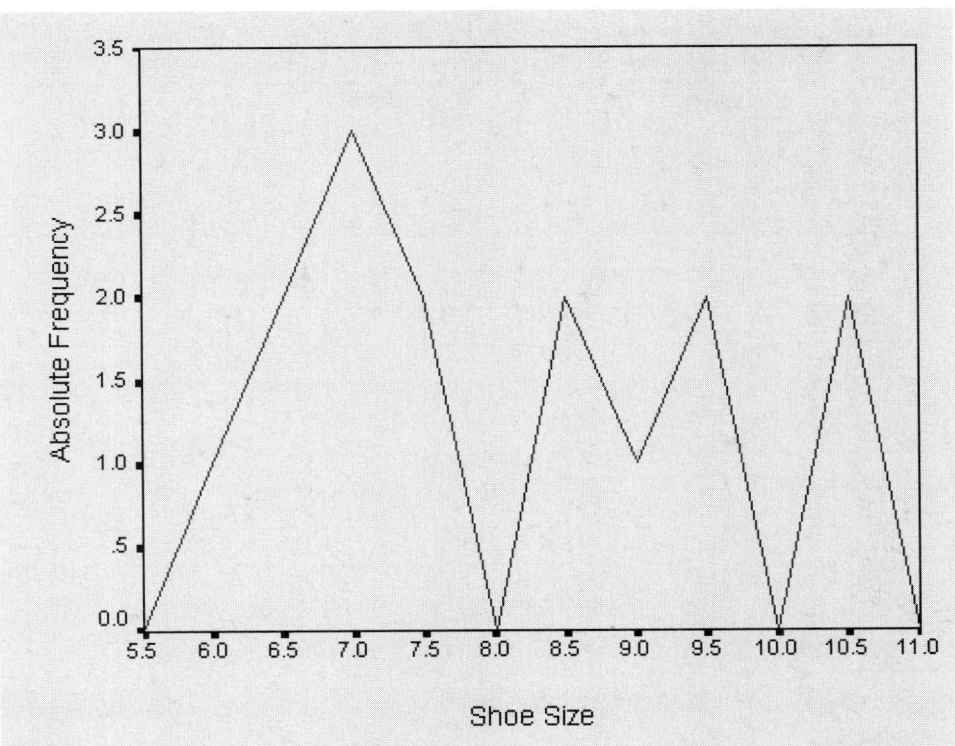

Absolute frequency polygon.

Note that when the frequency for a score is zero, as is the case for the shoe sizes of 8.0 and 10.0, the line goes down to the X-axis. Failing to go down to the X-axis when the frequency is zero is the most common error students make in drawing non-cumulative frequency polygons. This absolute frequency polygon was drawn using an indirect method in SPSS. A new data set was constructed from the frequency table as shown in the following figure:

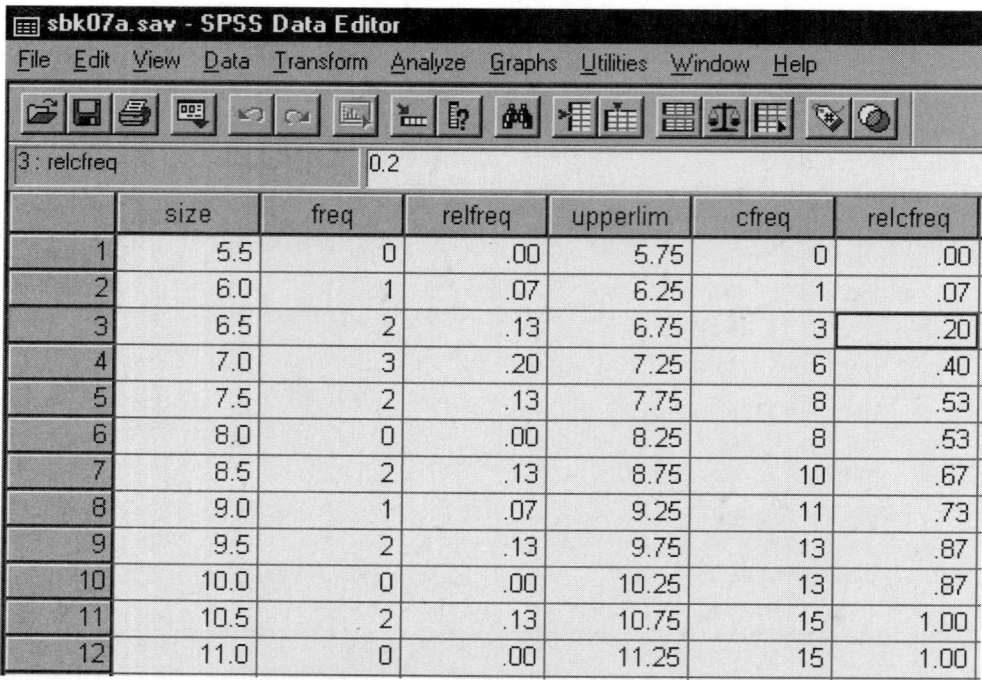

	size	freq	relfreq	upperlim	cfreq	relcfreq
1	5.5	0	.00	5.75	0	.00
2	6.0	1	.07	6.25	1	.07
3	6.5	2	.13	6.75	3	.20
4	7.0	3	.20	7.25	6	.40
5	7.5	2	.13	7.75	8	.53
6	8.0	0	.00	8.25	8	.53
7	8.5	2	.13	8.75	10	.67
8	9.0	1	.07	9.25	11	.73
9	9.5	2	.13	9.75	13	.87
10	10.0	0	.00	10.25	13	.87
11	10.5	2	.13	10.75	15	1.00
12	11.0	0	.00	11.25	15	1.00

Setting up the SPSS Data Editor to draw frequency polygons.

The graph is drawn by selecting Graphs/Line Charts. In the Line Charts dialog box that appears (see the following figure), select Simple, choose the Values of individual cases option, and click the Define button.

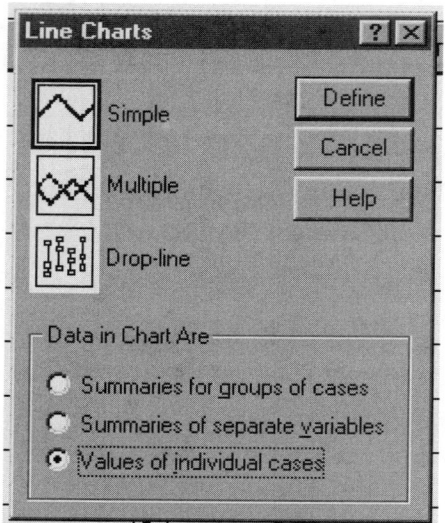

Selecting simple line charts using SPSS.

In the next screen (shown in the following figure), select the columns to use in the display. All of the subsequent graphs will be created in a similar manner by selecting different variables as rows and columns.

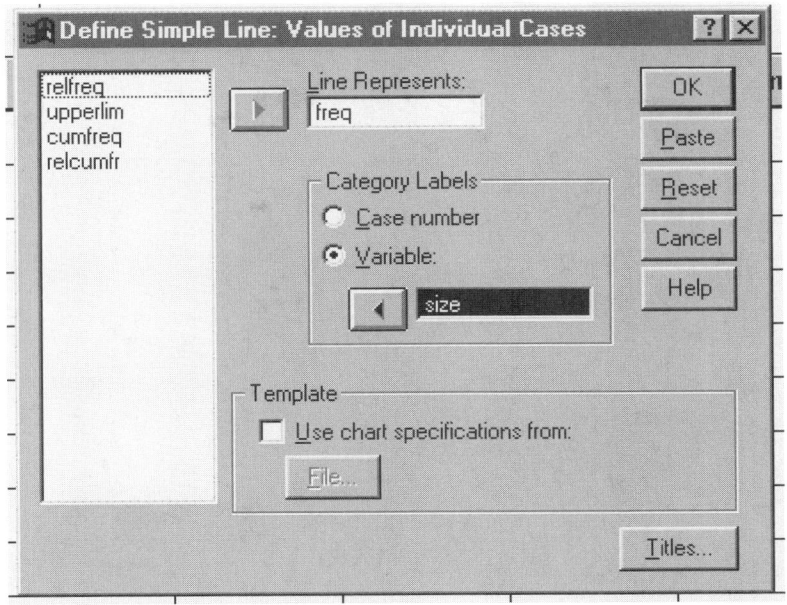

Defining the line in drawing frequency polygons using SPSS.

5-2c Relative Frequency Polygons

In order to draw a **relative frequency polygon,** the relative frequency of each score interval must first be calculated and placed in the appropriate column in the frequency table.

The **relative frequency** of a score is another name for the proportion of scores that have a particular value. The relative frequency is computed by dividing the frequency of a score by the number of scores (N). The following frequency table of the example data shows the additional column of relative frequencies.

Relative Frequency of Shoe Size		
Shoe Size	Absolute Frequency	Relative Frequency
6.0	1	.07
6.5	2	.13
7.0	3	.20
7.5	2	.13
8.0	0	.00
8.5	2	.13
9.0	1	.07
9.5	2	.13
10.0	0	.00
10.5	2	.13
	15	.99

The relative frequency polygon is drawn exactly like the absolute frequency polygon except the Y-axis is labeled and incremented with relative frequency rather than absolute frequency. The following frequency distribution is a relative frequency polygon. Note that it appears almost identical to the absolute frequency polygon.

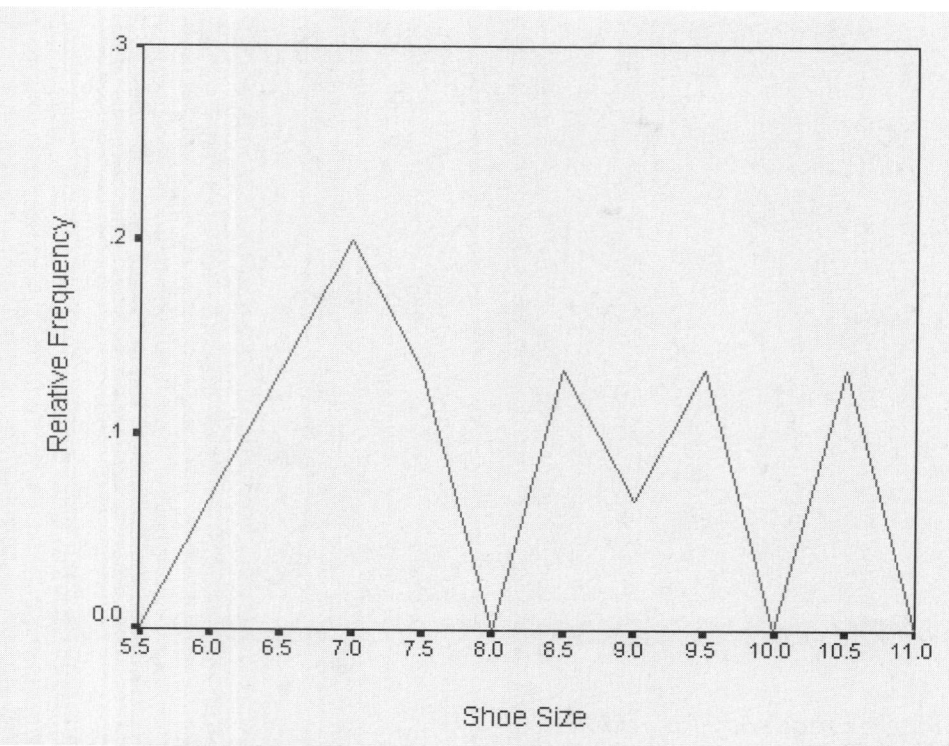

Resulting relative frequency polygon using SPSS.

A relative frequency may be transformed into an absolute frequency by using an opposite transformation; that is, multiplying by the number of scores (N). For this reason the size of the sample on which the relative frequency is based is usually presented somewhere on the graph. Generally speaking, relative frequency is more useful than **absolute frequency,** because the size of the sample has been taken into account.

5-2d Absolute Cumulative Frequency Polygons

An **absolute cumulative frequency** is the number of scores that fall at or below a given score value. It is computed by adding up the number of scores which are equal to or less than a given score value. The **cumulative frequency** may be found from the absolute frequency by either adding up the absolute frequencies of all scores smaller than or equal to the score of interest, or by adding the absolute frequency of a score value to the cumulative frequency of the score value immediately below it. The following frequency table includes a column for the absolute cumulative frequencies of the shoe size example data.

Shoe Size	Abs. Freq.	Abs. Cum. Freq.
6.0	1	1
6.5	2	3
7.0	3	6
7.5	2	8
8.0	0	8
8.5	2	10
9.0	1	11
9.5	2	13
10.0	0	13
10.5	2	15
	15	

Note that the cumulative frequency of the largest score (10.5) is equal to the number of scores (N = 15). This will always be the case if the cumulative frequency is computed correctly. The computation of the cumulative frequency for the score value of 7.5 could be done by either adding up the absolute frequencies for the scores of 7.5, 7.0, 6.5, and 6.0 (2 + 3 + 2 + 1 = 8), or adding the absolute frequency of 7.5, which is 2, to the absolute cumulative frequency of 7.0, which is 6, to get a value of 8.

Plotting scores on the X-axis and the absolute cumulative frequency on the Y-axis draws the cumulative frequency polygon. The points are plotted at the intersection of the upper real limit of the interval and the absolute cumulative frequency. The upper real limit is used in all cumulative frequency polygons because of the assumption that not all of the scores in an interval are accounted for until the upper real limit is reached. The following absolute cumulative frequency polygon presents the shoe size example data.

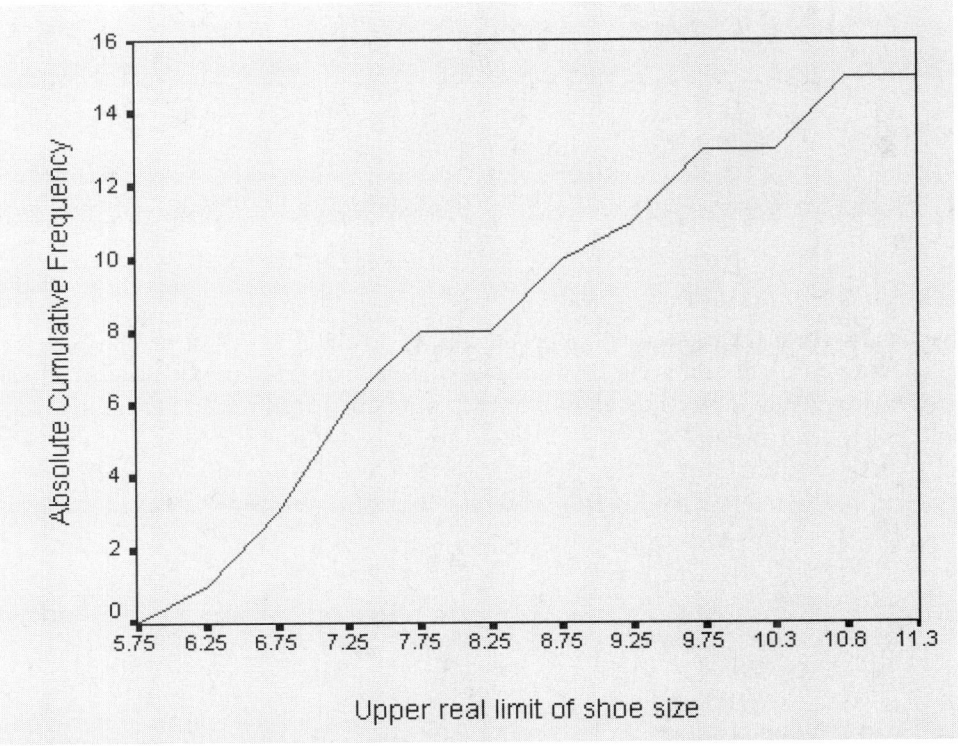

Example cumulative frequency polygon drawn using SPSS.

A **cumulative frequency polygon** will always be monotonically increasing, a mathematician's way of saying that the line will never go down, but that it will either stay at the same level or increase. The line will be horizontal when the absolute frequency of the score is zero, as is the case for the score value of 8.0 in the shoe size example. When the highest score is reached, 10.5 in this case, the line continues horizontally forever from that point. The cumulative frequency polygon, while displaying exactly the same amount of information as the absolute frequency distribution, expresses the information as a rate of change. The steeper the slope of the cumulative frequency polygon, the greater the rate of change. The slope of the example cumulative polygon is steepest between the values of 6.75 and 7.25, indicating the greatest number of scores falling between those values.

Rate of change information may be easier to comprehend if the score values involve a measure of time. The graphs of rate of rat bar pressing drawn by a behavioral psychologist are absolute cumulative polygons. Here's an example of a rat-press graph (Brandon, 2000):

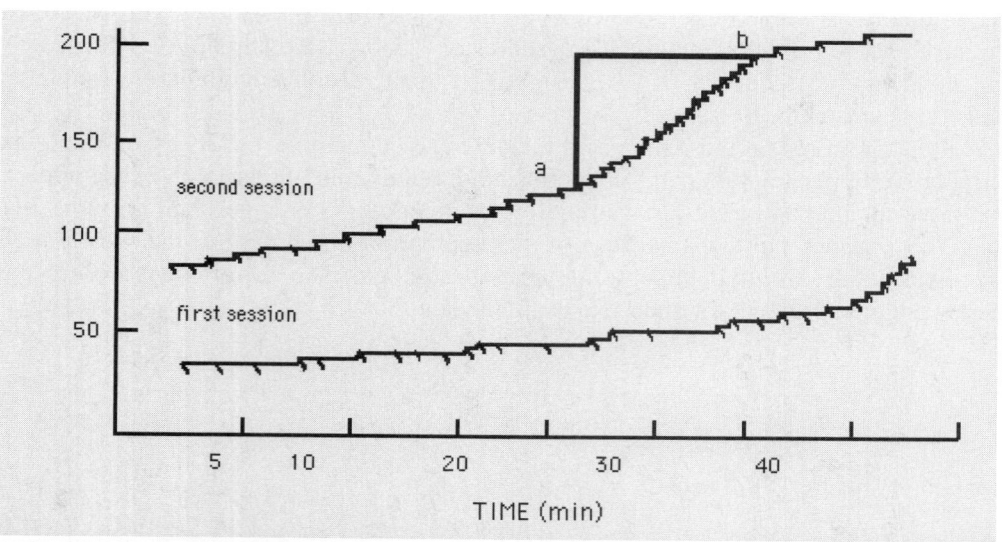

Cumulative record showing shaping and continuous reinforcement of the lever-pressing response by a rat.

5-2e Relative Cumulative Polygons

A **relative cumulative frequency** is the proportion of scores that fall at or below a given score value. It is computed by dividing the absolute cumulative frequency by the number of scores (N). The following frequency table shows the relative cumulative frequencies computed for the shoe size sample.

Shoe Size	Abs. Freq.	Abs. Cum. Freq	Rel. Cum. Freq.
6.0	1	1	.06
6.5	2	3	.20
7.0	3	6	.40
7.5	2	8	.53
8.0	0	8	.53
8.5	2	10	.67
9.0	1	11	.73
9.5	2	13	.87
10.0	0	13	.87
10.5	2	15	1.00
	15		

Drawing the X-axis as before and the relative cumulative frequency on the Y-axis draws the **relative cumulative frequency polygon** directly from the preceding table. Points are plotted at the intersection of the upper real limit and the relative cumulative frequency. The following graph shows the results from the shoe size example data.

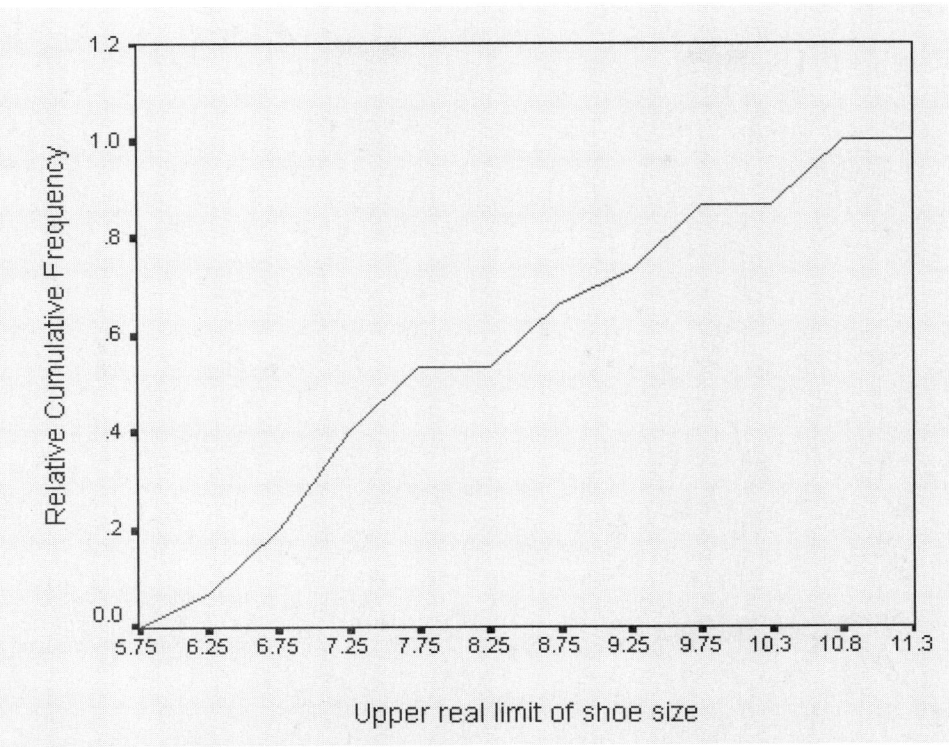

Example relative cumulative frequency polygon drawn using SPSS.

Note that the absolute and relative cumulative frequency polygons are identical except for the Y-axis. Note also that the value of 1.000 is the largest relative cumulative frequency, and the highest point on the polygon.

Summary

This chapter described how to draw pictures of data, which are called frequency distributions.

The first step in drawing a frequency distribution is to construct a frequency table. Then the information contained in the table may be transformed to a graphical or pictorial form. No information is gained or lost in this transformation, but people often find a graphical or pictorial presentation easier than tables of numbers to comprehend.

There are two major means of drawing a graph, histograms and frequency polygons. The choice of method is often a matter of convention, although there are times when one or the other is clearly the appropriate choice. You learned to draw histograms and to create several kinds of frequency polygons: relative frequency polygons, absolute frequency polygons, absolute cumulative frequency polygons, and relative cumulative polygons.

Chapter

6

Comparing Frequency Distributions

Key Terms

contingency tables
marginal frequencies
noncumulative polygons

overlapping frequency
 distributions
overlapping relative cumulative
 frequency polygons

overlapping relative frequency
 polygons
relative frequency
subsamples

When one variable is assumed to be measured on an interval scale, and another is dichotomous (that is, has only two levels), it is possible to illustrate the relationship between the variables by drawing **overlapping frequency distributions.** In the data presented in the preceding chapter shoe size could be treated as an interval measure and gender was a dichotomous variable with two levels, male and female. The relationship between gender and shoe size is thus an appropriate candidate for overlapping frequency distributions. Overlapping frequency distributions would be useful for two reasons: males wear different styles of shoes than females, and male and female shoe sizes are measured using different scales.

6-1 Overlapping Frequency Distributions

The first step in drawing overlapping frequency distributions is to partition the measured variable into two **subsamples,** one for each level of the dichotomous variable. In the example, shoe sizes are grouped into male and female groups as the following table shows:

Shoe Sizes for Males and Females								
Males	10.5	9.5	10.5	8.5	9.5	9.0		
Females	6.0	8.5	7.0	7.0	6.5	7.0	7.5	6.5	7.5

A separate frequency table is then computed for each subsample. The example results are shown in the following frequency table:

Frequency Table of Shoe Size for Males and Females				
	Males		**Females**	
Shoe Size	Abs Freq	Rel Freq	Abs Freq	Rel Freq
6.0	0	.00	1	.11
6.5	0	.00	2	.22
7.0	0	.00	3	.33
7.5	0	.00	2	.22
8.0	0	.00	0	.00
8.5	1	.17	1	.11
9.0	1	.17	0	.00
9.5	2	.33	0	.00
10.0	0	.00	0	.00
10.5	2	.33	0	.00
	6	1.00	9	.99

Note that the **relative frequency** is computed by dividing the absolute frequency by the number of scores in that group. For example, the relative frequency of shoe size 9.5 for males is 2 (the number of males wearing a size 9.5 shoe) divided by 6 (the total number of males). The sum of the relative frequency for each gender must equal 1.00, within rounding error. In this case, by rounding the relative frequencies to two decimal places, the sum of the relative frequencies will most likely be between .98 and 1.02. Any sum outside this range is most likely due to an error in computation and not rounding error.

6-2 Overlapping Relative Frequency Polygons

To draw **overlapping relative frequency polygons** using SPSS/WIN, enter the relative frequency table as data. The example data has been entered in the following figure:

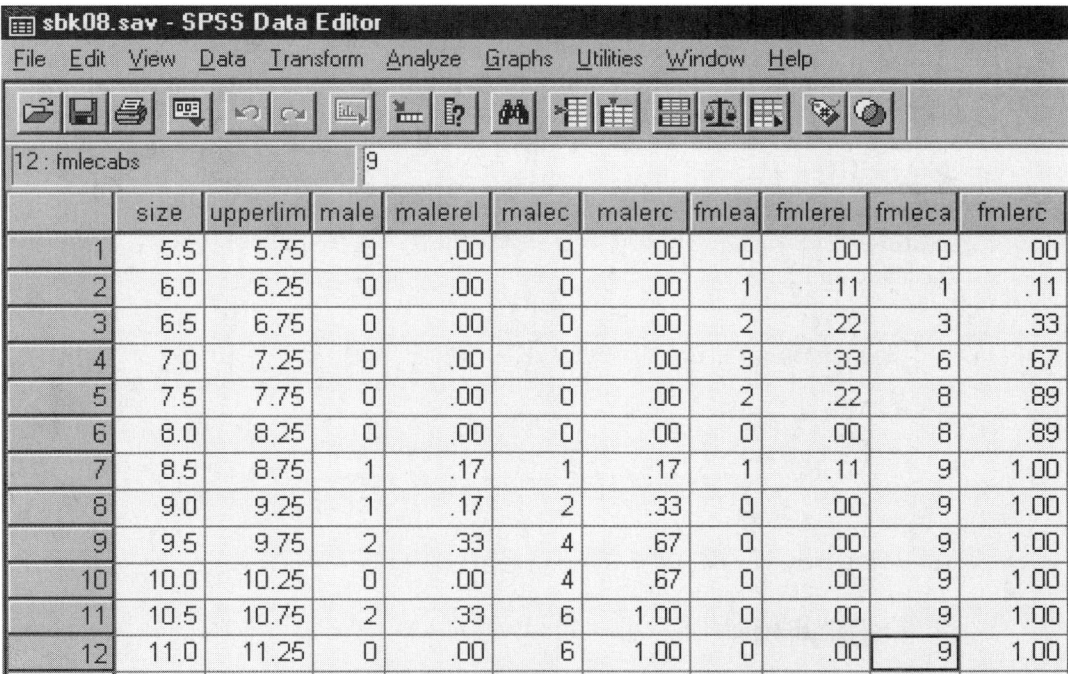

Data Editor in SPSS for overlapping relative frequency polygons.

The overlapping relative frequency polygons are simply the two polygons for each group drawn on the same set of axes, distinguished with different types of lines. If conflicts appear, they may be resolved by drawing the lines next to one another. Here's an example of overlapping relative frequency polygons:

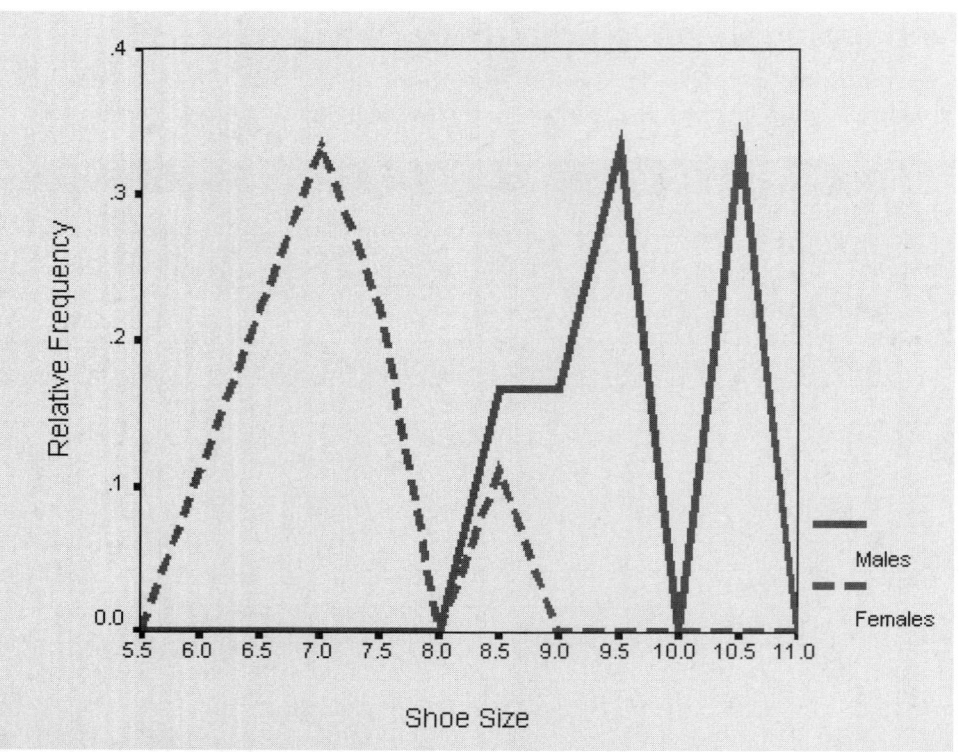

Overlapping relative frequency polygons.

When polygons are drawn in this form, they may be easily compared with respect to their centers, shapes, continuity, and so forth.

Overlapping relative cumulative frequency polygons may also give additional information about how two distributions are similar or different. In many ways these polygons are easier to interpret because the lines do not jump up and down as much as in the **noncumulative polygons.** The following graph illustrates two overlapping relative cumulative frequency polygons of shoe size for males and females.

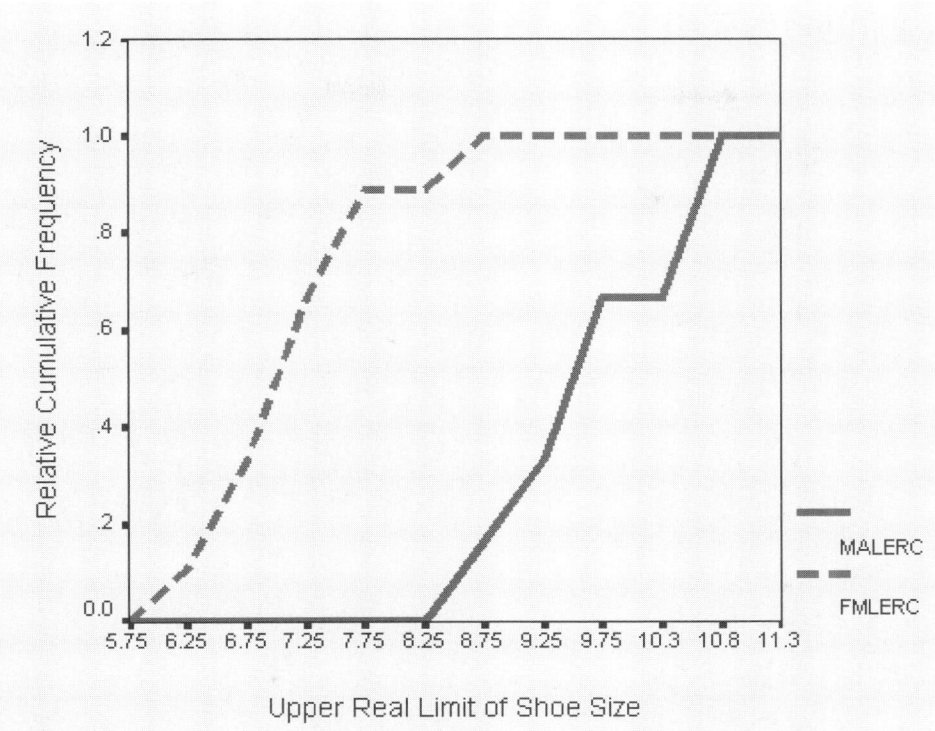

Overlapping cumulative frequency polygons of shoe sizes for males and females.

The procedure to construct the preceding graphs in SPSS/WIN is first to enter a frequency table as previously described and then to select the Graphs/Line/Multiple options from the toolbar. A multiline graph will generate the desired results. The commands necessary to generate the overlapping relative cumulative frequency polygons are illustrated in the following figure. After you choose your options, click OK.

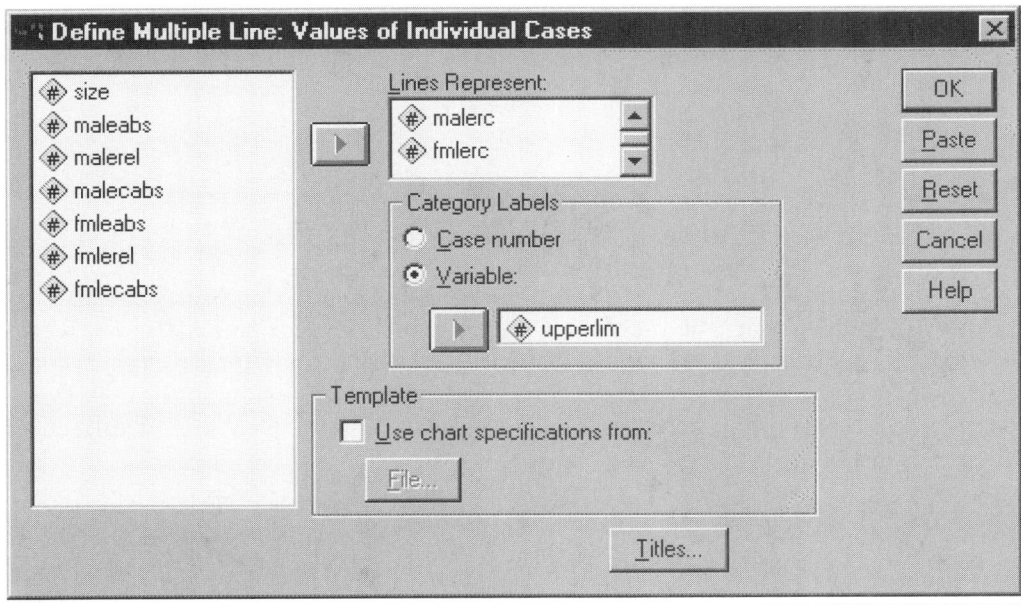

SPSS drop-down box to draw overlapping relative frequency polygons.

6-3 Contingency Tables

Frequency tables of two variables presented simultaneously are called **contingency tables.** Although this rule is sometimes broken, contingency tables are generally appropriate for variables that have five or fewer levels, or different values. More than five levels, while not technologically incorrect, may result in tables that are very difficult to read and should be used with caution.

Contingency tables are constructed by listing all the levels of one variable as rows in a table and the levels of the other variables as columns. For example, the labeling of the contingency table of gender by shoe width is illustrated here:

		Shoe	Width		
		A	B	C	D
Sex	Male				
	Female				

Structure of a contingency table.

The second step in computing the contingency table is to find the joint or cell frequency for each cell. For example, the cell in the upper left corner contains the number of males who had shoe width of A, which in this case is zero. In turn, each cell has its frequency counted and placed in the appropriate cell. The cell frequencies are then summed both across the rows and down the columns. The sums are placed in the margins, the values of which are called **marginal frequencies.** The lower right hand corner value contains the sum of either the row or column marginal frequencies, which must match and be equal to N. At this point, the contingency table looks like this:

	Shoe	Width			
	A	B	C	D	
Male	0	1	2	3	6
Female	2	6	1	0	9
	2	7	3	3	15

Example contingency table with absolute frequencies.

This absolute frequency table may be converted to a relative frequency table by dividing the absolute cell frequency by the number of scores, which may be row marginal frequencies, column marginal frequencies, or overall frequency (N). In the previous example, computing relative frequencies with respect to the row marginal frequencies results in the following table. This table gives the proportion of males or females who have a given shoe width, and would probably be most useful in ordering shoes.

	Shoe Width				
	A	B	C	D	
Male	0 / .00	1 / .17	2 / .33	3 / .50	6
Female	2 / .22	6 / .67	1 / .11	0 / .00	9
	2	7	3	3	15

Example contingency table with relative column frequencies.

Computing the cell proportions using the column marginal frequencies, expressing the proportion of each shoe width which was male or female, is probably not as useful, but is shown below as a second possibility.

	Shoe Width				
	A	B	C	D	
Male	0 / .00	1 / .14	2 / .67	3 / 1.00	6
Female	2 / 1.00	6 / .86	1 / .33	0 / .00	9
	2	7	3	3	15

Example contingency table with relative row frequencies.

Summary

This chapter showed you how to illustrate the relationship between the variables by drawing overlapping frequency distributions and by creating contingency tables.

In drawing overlapping frequency distributions, you learned to partition the measured variable into two subsamples, one for each level of the dichotomous variable. Overlapping relative frequency polygons are simply the polygons for both groups drawn on the same set of axes, distinguished with different types of lines.

Frequency tables of two variables presented simultaneously are called contingency tables. Contingency tables are constructed by listing all the levels of one variable as rows in a table and the levels of the other variables as columns. Contingency tables are a convenient means of showing the relationship between two variables. When relative frequencies are computed, useful information about the distribution of a single variable over levels of another variable may be presented.

Chapter

7

Grouped
Frequency Distributions

Key Terms

apparent limits   grouped frequency polygons   real limits

An investigator interested in finger-tapping behavior conducts the following study: Students are asked to tap as fast as they can with their ring finger. The hand is cupped and all fingers except the one being tapped are placed on the surface. Either the right or the left hand is used, at the preference of the student. At the end of 15 seconds, the number of taps for each student is recorded. Example data using 18 subjects are presented here:

Finger Taps in Fifteen Seconds for Eighteen Individuals																	
53	35	67	48	63	42	48	55	33	50	46	45	59	40	47	51	66	53

The following figure shows part of the data file in SPSS corresponding to the example data:

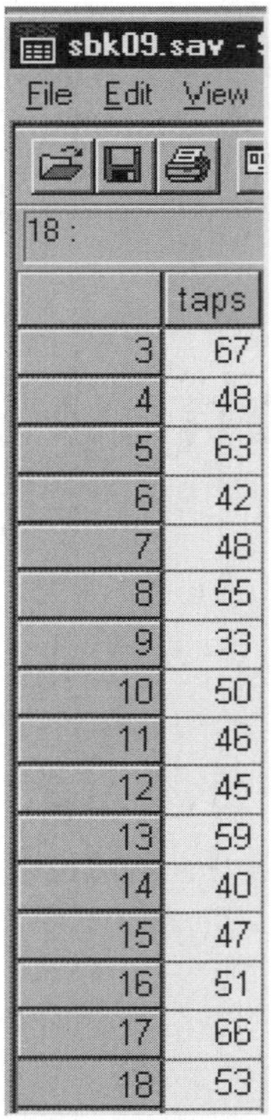
sbk09.sav -
File Edit View
18 :
taps
3 67
4 48
5 63
6 42
7 48
8 55
9 33
10 50
11 46
12 45
13 59
14 40
15 47
16 51
17 66
18 53

Example data file for grouped frequency polygons in SPSS.

The frequency table resulting from this data would have 34 different score values, computed by subtracting the low score (33) from the high score (67). A portion of this table is shown here.

Absolute Frequency Table of Ungrouped Data

# Taps	Absolute Frequency
33	1
34	0
35	1
...	...
65	0
66	1
67	1
	18

The following figure shows a histogram drawn using this data:

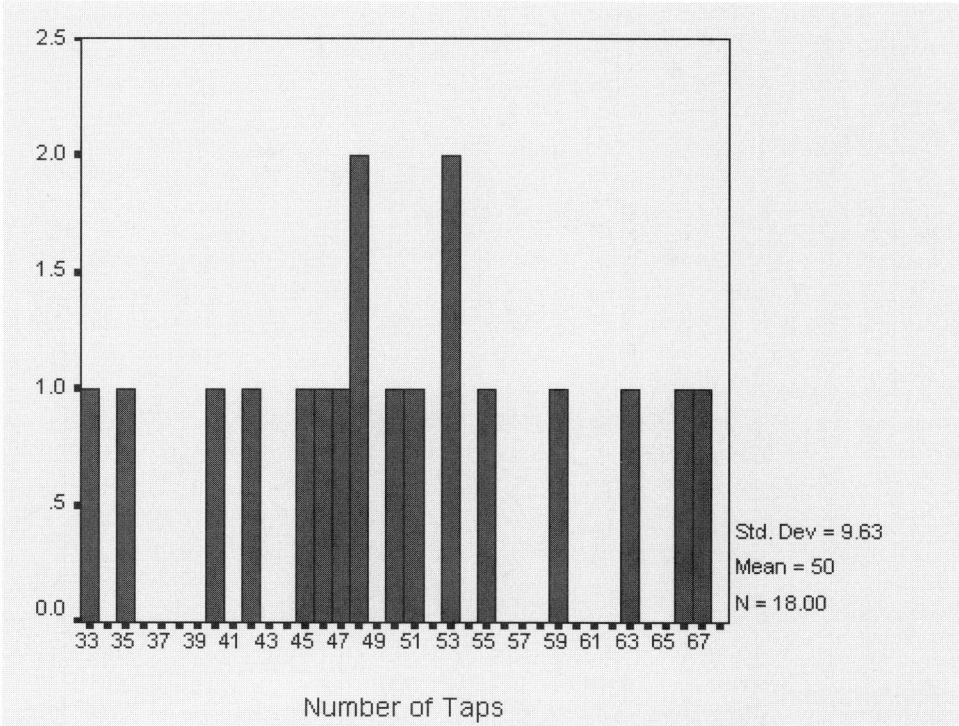

Example of a sawtoothed histogram with ungrouped data.

7-1 Why Intervals Are Necessary

The previous table and graph present all the information possible given the data. The problem is that so much information is presented that it is difficult to discern what the data is really like, or to "cognitively digest" the data. The graph is given the term "sawtoothed" because the many ups and downs give it the appearance of teeth on a saw. The great amount of data ink relative to the amount of information on the polygon makes an alternative approach desirable. It is possible to lose information (precision) about the data to gain understanding about distributions. This is the function of grouping data into intervals and drawing **grouped frequency polygons.**

The process of drawing grouped frequency distributions can be broken down into a number of interrelated steps: selecting the interval size, computing the frequency table, and drawing the grouped frequency histogram or polygon. Each will now be discussed in turn.

7-2 Selecting the Interval Size

The goal in selecting an interval size is to present as much information to the reader as possible in the simplest form possible. In some cases, as illustrated in the previous example, the graph contains too much information to be easily readable. In grouping the data into intervals, the statistician reduces the information in the graph in order to make it easier to understand. There is a trade-off, then, between the amount of information in the graph and the difficulty in reading the graph. The statistician desires to select the point at which the graph containing the greatest amount of information is presented in a readable form. This point may be different depending upon the reader.

Selecting the interval size is more art than science. The statistician has to begin somewhere, so the following procedure is fairly standard. To determine a starting interval size, the first step is to find the range of the data by subtracting the smallest score from the largest. In the case of the example data, the range was 67 − 33 = 34. The range is then divided by the number of desired intervals, with a suggested starting number of intervals being ten (10). In the example, the result would be 34/10 = 3.4. The nearest odd integer value is used as the starting point for the selection of the interval size, so for our example, that would be 3.

7-3 Computing the Frequency Table

After the interval size has been selected, the scale is then grouped into equal-sized intervals based on the interval size. The first interval will begin with a multiple of the interval size equal to, or smaller than, the smallest score. In the example the first interval would begin with the value of 33, a multiple of the interval size (3 * 11). In this case the beginning of the first interval equals the smallest score value.

The ending value of the first interval is computed by adding the interval size to the beginning of the first interval and subtracting the unit of measurement. In the example, the beginning of the first interval (33) plus the interval size (3) minus the unit of measurement (1) results in a value of 33 + 3 − 1 or 35. Thus the first interval would be 33 to 35. Sequentially adding the interval size to these values results in all other intervals, for example 36 to 38, 39 to 41, and so on.

The values for the intervals just constructed are called the **apparent limits** of the intervals. In the first interval, for example, the value of 33 would be called the apparent lower limit, and the value of 35 would be the apparent upper limit.

The midpoints of the intervals are computed by adding the two apparent limits together and dividing by two. The midpoint for the interval 33 to35 would thus be (33 + 35)/2 or 34. The midpoint for the second interval (36 − 38) would be 37.

The midpoints between midpoints are called **real limits.** Each interval has a real lower limit and a real upper limit. The interval 36 − 38 would therefore have a real lower limit of 35.5 and a real upper limit of 38.5. Please note that the difference between the real limits of an interval is equal to the interval size, that is 38.5 − 35.5 = 3. All this is easier than it first appears, as can be seen in the following table:

Apparent and Real Limits for an Interval Size of Three					
	Apparent		Real		
Interval	Lower Limit	Upper Limit	Lower Limit	Upper Limit	Midpoint
33–35	33	35	32.5	35.5	34
36–38	36	38	35.5	38.5	37
39–41	39	41	38.5	41.5	40
42–44	42	44	41.5	44.5	43
45–47	45	47	44.5	47.5	46
48–50	48	50	47.5	50.5	49
51–53	51	53	50.5	53.5	52
54–56	54	56	53.5	56.5	55
57–59	57	59	56.5	59.5	58
60–62	60	62	59.5	62.5	61
63–65	63	65	62.5	65.5	64
66–68	66	68	65.5	68.5	67

The hard work is finished when the intervals have been selected. All that remains is the counting of the frequency of scores for each interval, and, if needed, computing the relative, cumulative, and relative cumulative frequencies for the intervals. The frequency table for the example data with an interval size of three is presented here:

Absolute Frequency Table for an Interval Size of Three	
Interval	Absolute Frequency
33–35	2
36–38	0
39–41	1
42–44	1
45–47	3
48–50	3
51–53	3
54–56	1
57–59	1
60–62	0
63–65	1
66–68	2

7-4 Drawing the Frequency Polygon or Histogram

The frequency histogram or polygon is drawn by plotting the midpoints of the intervals on the x-axis and the frequency on the y-axis. The following is an absolute frequency polygon of the example data:

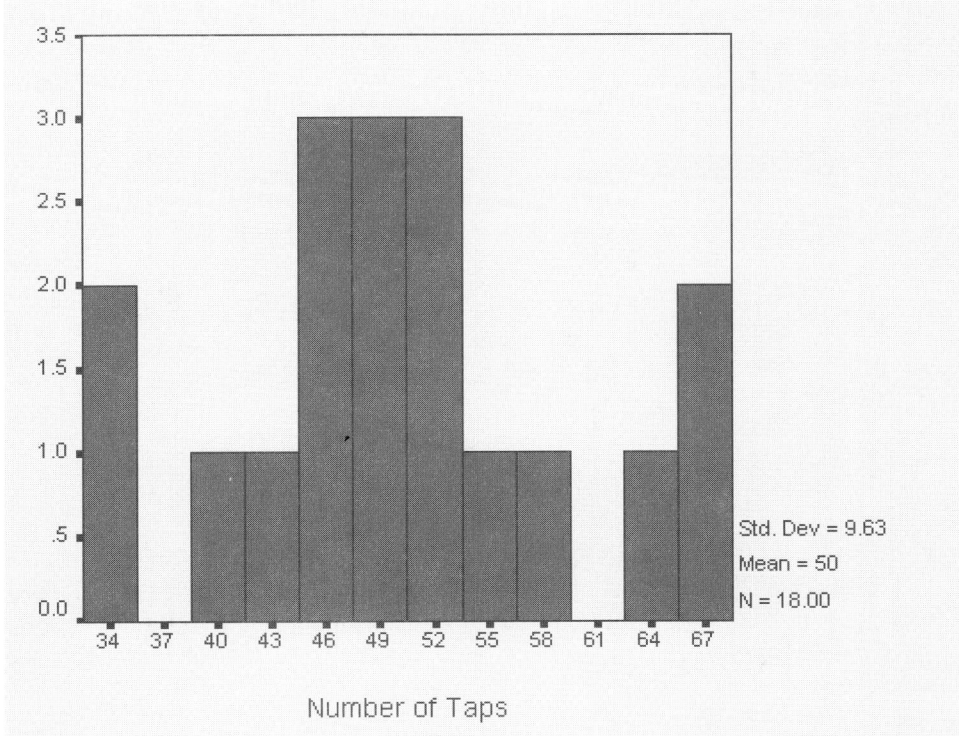

Histogram of example data grouped into intervals of 3.

This histogram was generated using SPSS graphs commands. The graph was first generated by selecting Graphs/Histogram. In order to select the appropriate interval, the resulting image was edited and the category axis was changed as the following figure shows:

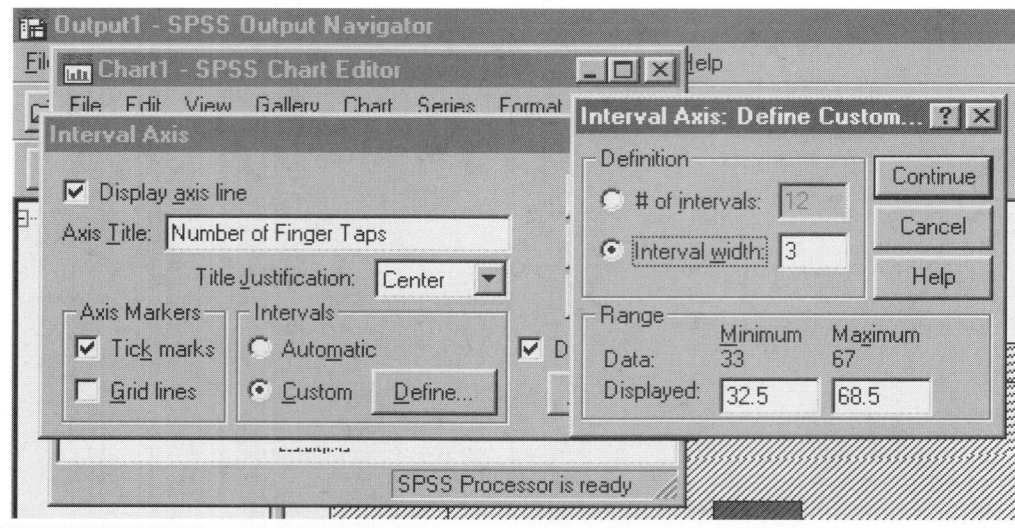

SPSS command to change the interval size of the histogram.

All of the histograms presented in this chapter were generated in a similar manner. Selecting the appropriate interval size and real lower limit will produce the desired result.

7-5 Selecting Another Interval Size

The first interval selected might not be the interval that best expresses or illustrates the data. A larger interval will condense and simplify the data; a smaller interval will expand the data and make the picture more detailed. Here is an alternative frequency table for the example data with an interval of 6:

		Apparent and Real Limits for an Interval of Six				
		Apparent		Real		
Interval	Lower Limit	Upper Limit	Lower Limit	Upper Limit	Midpoint	Abs. Freq.
30–35	30	35	29.5	35.5	32.5	2
36–41	36	41	35.5	41.5	38.5	1
42–47	42	47	41.5	47.5	44.5	4
48–53	48	53	47.5	53.5	50.5	6
54–59	54	59	53.5	59.5	56.6	2
60–65	60	65	59.5	65.5	62.5	1
66–71	66	71	65.5	71.5	68.5	2
						18

Note that for the first interval, the apparent lower limit is 30, the apparent upper limit is 35, the real lower limit is 29.5, the real upper limit is 35.5, and the midpoint is 32.5. The midpoint is not a unit of measurement, like 33, but a half unit, 32.5. The problem of having a midpoint that is not a unit of measurement is due to the even interval size, six in this case. For this reason, odd interval sizes are preferred.

The following histogram represents the frequency table using intervals of 6:

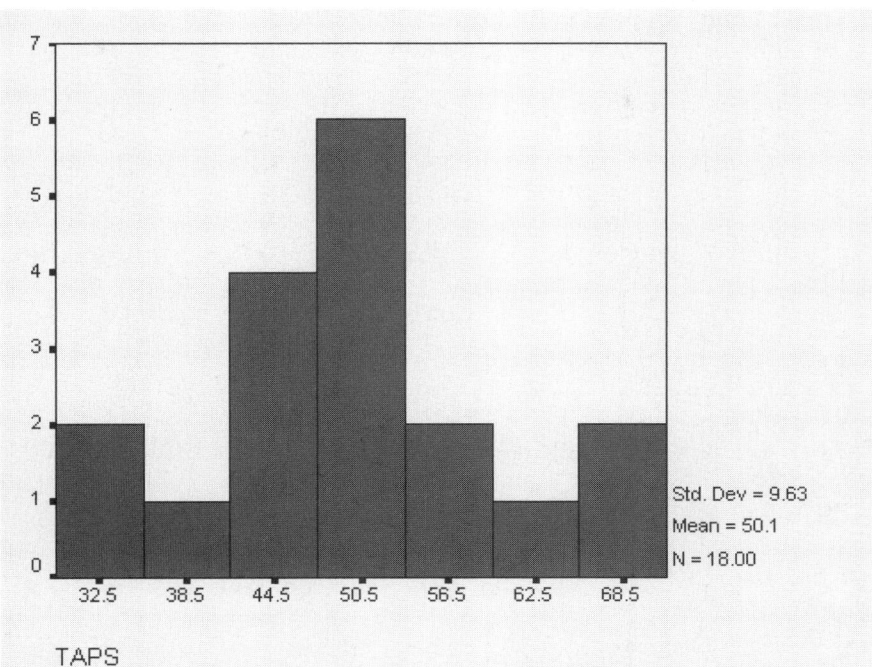

Grouped frequency histogram with interval of 6.

7-6 Selecting the Appropriate Interval Size

Selection of the appropriate interval size requires that the intended audience of the graph be constantly kept in mind. If the persons reading the graph are likely to give the picture a cursory glance, then the information must be condensed by selecting a larger interval size. If detailed information is necessary, then a smaller interval size must be selected. The selection of the interval size, therefore, is a trade-off between the amount of information present in the graph, and the difficulty of reading the information.

Factors other than the interval size, such as the number of scores and the nature of the data, also affect the difficulty of the graph. Because of this, my recommendation is to select more than one interval size, draw the associated polygons, and use the resulting graph which best expresses the data for the purposes of the given audience. In this case there are no absolutes in drawing frequency polygons.

The frequency table and resulting histogram for the example data with an interval of 5 follow:

	Apparent and Real Limits for an Interval Size of Five					
	Apparent		Real			
Interval	Lower Limit	Upper Limit	Lower Limit	Upper Limit	Midpoint	Abs. Freq.
30–34	30	34	29.5	34.5	32	1
35–39	35	39	34.5	39.5	37	1
40–44	40	44	39.5	44.5	42	2
45–49	45	49	44.5	49.5	47	5
50–54	50	54	49.5	54.5	52	4
55–59	55	59	54.5	59.5	57	2
60–64	60	64	59.5	64.5	62	1
65–69	65	69	64.5	69.5	67	2
						18

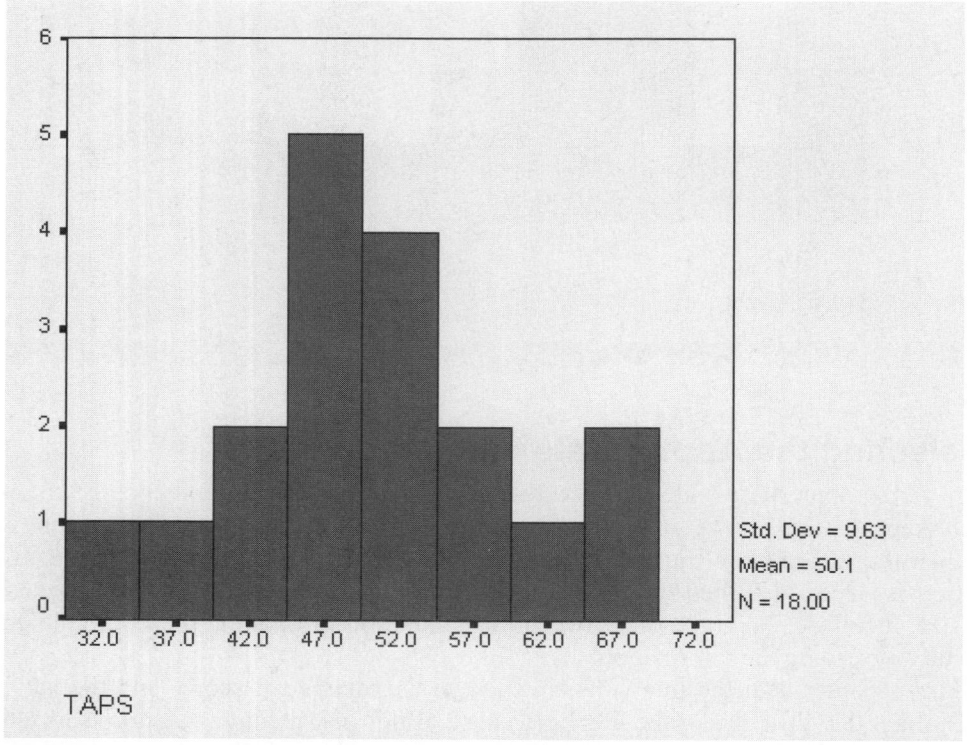

Grouped histogram with interval of 5.

And, so you can see how various interval sizes affect a graph, here are the histograms for the example data with intervals of 7, 9, and 11:

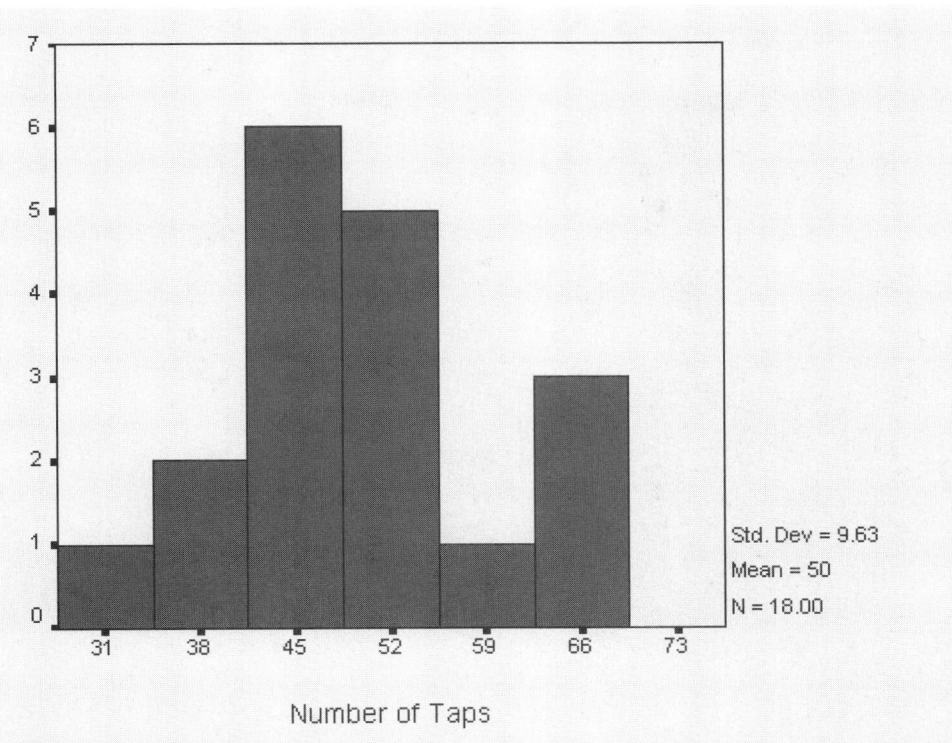

Std. Dev = 9.63
Mean = 50
N = 18.00

Number of Taps

Histogram with an interval of 7.

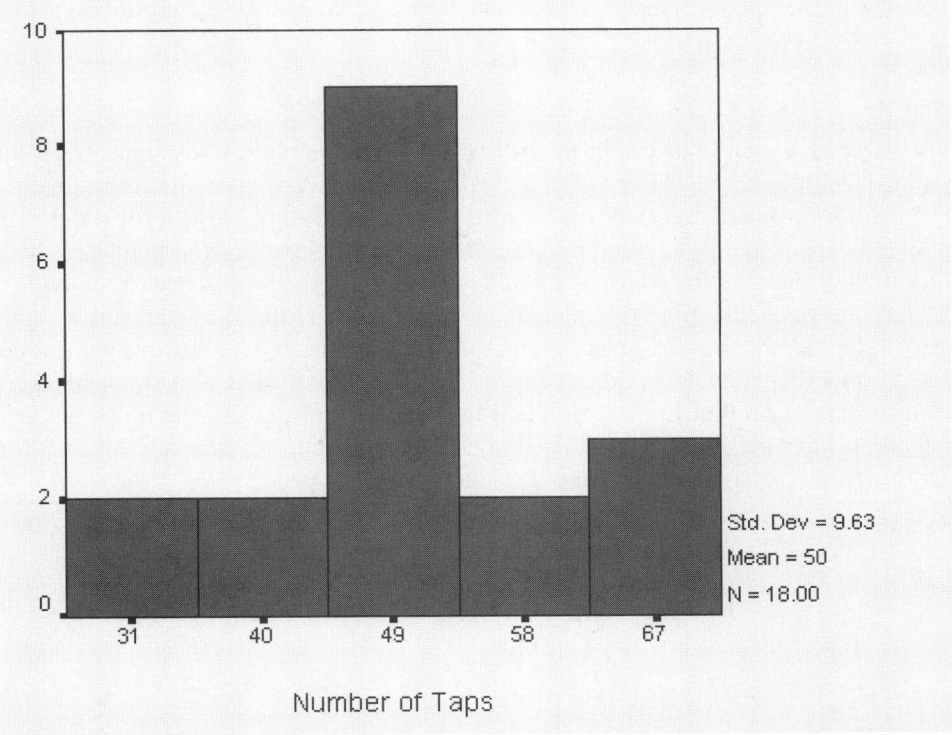

Std. Dev = 9.63
Mean = 50
N = 18.00

Number of Taps

Histogram with an interval of 9.

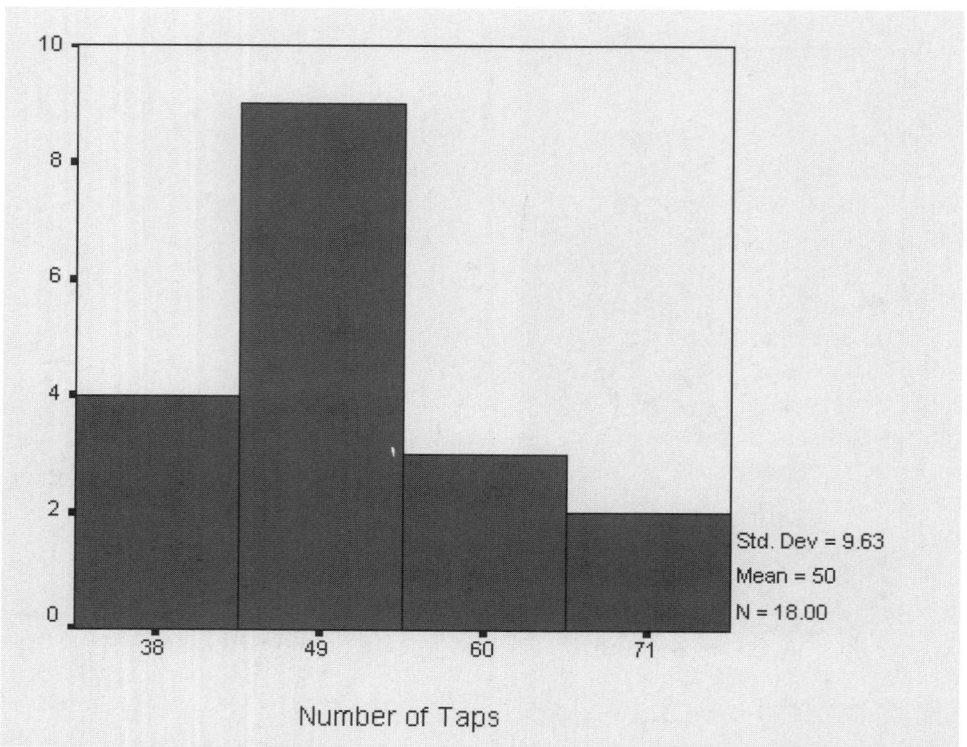

Std. Dev = 9.63
Mean = 50
N = 18.00

Number of Taps

Histogram with an interval of 11.

As can be seen, the shape of the distribution changes as different interval sizes are selected.

Summary

Oftentimes a frequency table or graph presenting all the information possible, given the data, is difficult for your audience to understand. A desirable alternative approach is to group data into intervals and draw grouped frequency polygons. You may lose a little information (precision) about the date, but you will gain greater comprehension of the distribution, which is the benefit of grouped frequency distributions. This chapter led you through the process of drawing grouped frequency distributions: selecting the interval size, computing the frequency table, and drawing the grouped frequency histogram or polygon. You learned about apparent limits, midpoints, and real limits of intervals, and how the selection of the interval size is a trade-off between the amount of information present in the graph and the difficulty in reading the information. You also saw how factors other than the interval size, such as the number of scores and the nature of the data, affect the difficulty of the graph.

Chapter

8

Models of Distributions

Key Terms

distribution
model of sample frequency distribution
negative exponential distribution
normal curve
normal distribution
parameters
population
probability
probability density function (pdf)
probability distribution
probability model
relative frequency
theoretical probability distribution
uniform distribution

A model of a frequency distribution is an algebraic expression describing the **relative frequency** (height of the curve) for every possible score. The questions that may come to mind are, "What is the advantage of this level of abstraction? Why is all this necessary?" The answers to these questions may be found in the following.

- **A belief in the eminent simplicity of the world**—The belief that a few algebraic expressions can adequately model a large number of very different real-world phenomena underlies much statistical thought. The fact that these models often work justifies this philosophical approach.

- **Not all the data can be collected**—In almost all cases in the social sciences, it is not feasible to collect data on the entire population in which the researcher is interested. For instance, the individual opening a shoe store might want to know the shoe sizes of all persons living within a certain distance of the store, but not be able to collect all the data. If data from a subset or sample of the population of interest were used, rather than the entire population, then repeating the data collection procedure would most likely result in a different set of numbers. A model of the **distribution** is used to give some consistency to the results.

For example, suppose that the distribution of shoe sizes collected from a sample of fifteen individuals resulted in the following relative frequency polygon.

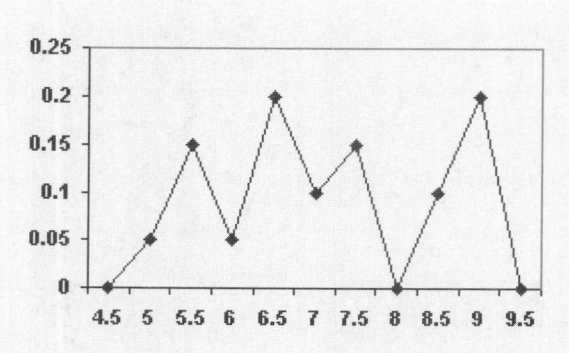

Relative frequency polygon of shoe sizes.

Since there are no individuals in the sample who wear size eight shoes, does that mean that the store owner should not stock the shelves with any size eight shoes? If a different sample were taken, would an individual who wore a size eight likely be included? Because it can reasonably be assumed that the reason no size eights were found in the sample was because of chance or sampling error, some method of ordering shoes other than directly from the sample distribution must be used.

In order to better deal with random fluctuations when collecting information from a sample, the statistician has the option of creating a **model of sample frequency distribution.** This model is called by different names, including **probability model, theoretical probability distribution, probability density function (pdf),** or simply **population.** A probability model attempts to capture the essential structure of the real world by asking what the world might look like if an infinite number of scores were obtained and each score was measured infinitely precisely. Nothing in the real world is exactly distributed as any given probability model. However, a probability model often describes the world well enough to be useful in making decisions.

For example, suppose that the frequency polygon of shoe size for women actually looked like the following:

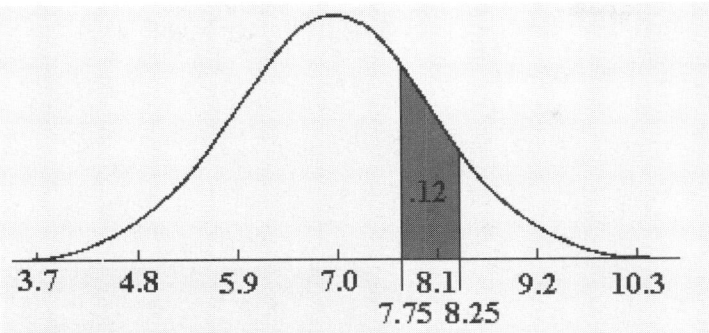

Area under the probability model between scores of 7.75 and 8.25.

If this were the case, the proportion (.12) or percentage (12%) of size eight shoes could be computed by finding the relative area between the real limits for a size eight shoe (7.75 to 8.25). The relative area between scores on any probability model is called **probability.** In this case, the probability of a randomly selected woman wearing a size eight shoe would be .12. The concept of area under a curve will be covered in more detail in a later chapter.

8-1 Variations of Probability Models

The statistician has at his or her disposal a number of probability models to describe the world. Different models are selected for practical or theoretical reasons. Some examples of probability models follow.

8-1a The Uniform or Rectangular Distribution

The **uniform distribution** is shaped like a rectangle, where each score is equally likely. The following figure shows an example:

The uniform distribution.

If the uniform distribution was used to model shoe size, it would mean that between the two extremes, each shoe size would be equally likely. If the store owner were ordering shoes, it would mean that an equal number of each shoe size would be ordered. In most cases this would be a very poor model of the real world, because at the end of the year a large number of large and small shoe sizes would remain on the shelves and the middle sizes would be sold out.

The uniform distribution is a useful model when the phenomena being modeled is relatively stable over a range of values. For example, the relative frequency of births on any day of the year in United States is relatively constant. In that case a uniform distribution might be an adequate, but not perfect, model.

8-1b The Negative Exponential Distribution

The **negative exponential distribution** is often used to model real-world events that are relatively rare, such as the occurrence of earthquakes. The negative exponential distribution would be a good model of the relative frequency of lottery winnings. Here's an example of an overly optimistic distribution:

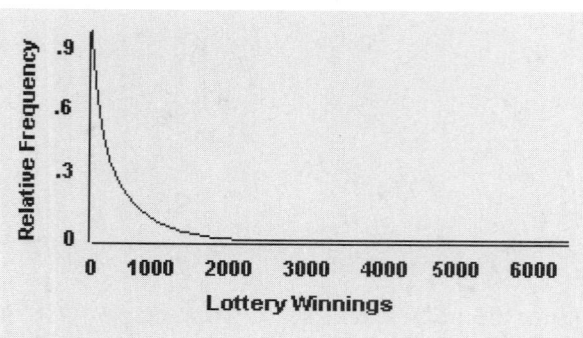

A negative exponential distribution.

8-1c The Triangular Distribution

Not really a standard distribution, a triangular distribution could be created as follows:

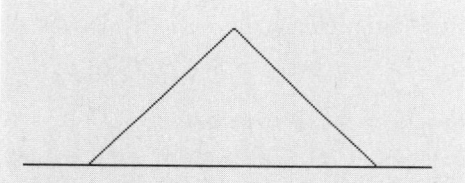

A triangular distribution.

It may be useful for describing some real-world phenomena, but exactly what that would be is not known for sure. The statistician has the option of creating a distribution for a particular situation if mathematical equations can be found to describe the model. The statistician is not limited to only distributions that are widely used or that others have already discovered.

8-1d The Normal Distribution or Normal Curve

The **normal distribution** or **normal curve** is one of a large number of possible distributions. It is very important in the social sciences and will be described in detail in the next chapter. The shoe size model presented earlier was an example of a normal curve, and here's another:

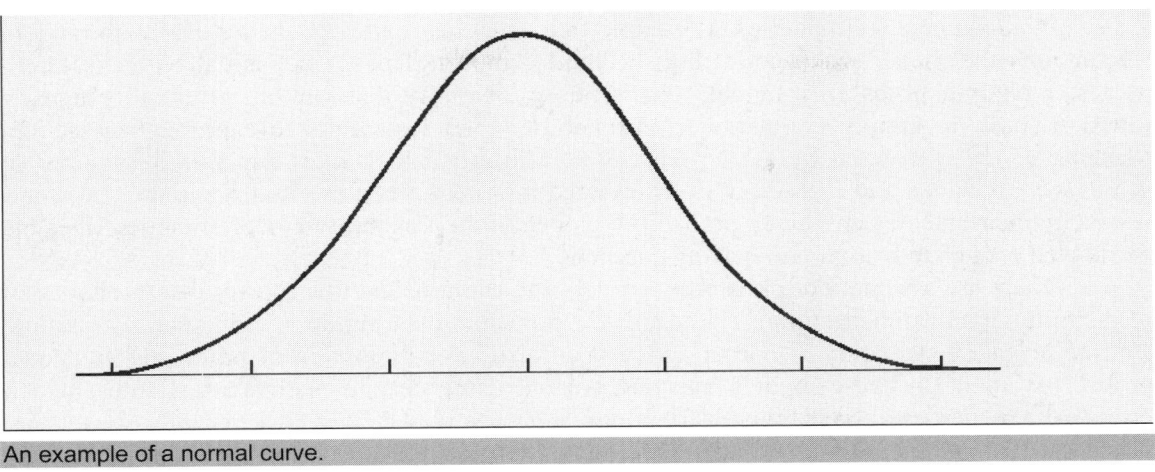
An example of a normal curve.

8-2 Properties of Probability Distributions

As mentioned earlier, the statistician has the option of producing his or her own probability models, which must be created using certain mathematical rules. These rules provide the properties of **probability distributions:**

- **Parameters**—Almost all of the useful models contain parameters. As you recall from Chapter 2, "Models," **parameters** are variables within the model that must be set before the model is completely specified. Changing the parameters of a probability model changes the shape of the curve. The use of parameters allows a single general-purpose model to describe a wide variety of real-world phenomena.

- **The Area underneath the Curve Is Equal to One (1.0)**—In order for an algebraic expression to qualify as a legitimate model of a distribution, the total area under the curve must be equal to one. This property is necessary for the same reason that the sum of the relative frequencies of a sample frequency table was equal to one. This property must hold true no matter what values are selected for the parameters of the model.

- **The Area under the Curve between Any Two Scores Is a Probability**—Although probability is a common term in the natural language, meaning likelihood or chance of occurrence, statisticians define it much more precisely. The probability of an event is the theoretical relative frequency of the event in a model of the population.

The models that have been discussed up to this point assume continuous measurement. That is, every score on the continuum of scores is possible, or there are an infinite number of scores. In this case, no single score can have a relative frequency because if it did, the total area would necessarily be greater than one. For that reason probability is defined over a range of scores rather than a single score. Thus a shoe size of 8.00 would not have a specific probability associated with it, although the interval of shoe sizes between 7.75 and 8.25 would.

Summary

In this chapter, you learned about models of distributions. A model of a frequency distribution is an algebraic expression describing the relative frequency (height of the curve) for every possible score. These models have been proven effective for two reasons: a belief in the eminent simplicity of the world, and the fact that not all the data can be collected.

You can create a model of the sample frequency distribution to better deal with random fluctuations that occur when collecting different samples. This model is called by different names, including probability model, theoretical probability distribution, probability density function (pdf), or simply population. A probability model attempts to capture the essential structure of the real world by asking what the world might look like if an infinite number of scores were obtained and each score was measured infinitely precisely. Nothing in the real world is exactly distributed as any given probability model, but a probability model often describes the world well enough to be useful in making decisions.

There are a number of probability models you can use, and this chapter discussed several of them: the uniform distribution, the negative exponential distribution, a triangular distribution, and the normal distribution or normal curve. You also have the option of producing your own probability models. These models must be created using certain mathematical rules, which provide the properties of probability distributions.

Statisticians define probability precisely: The probability of an event is the theoretical relative frequency of the event in a model of the population. The theoretical relative frequency or probability is defined as the area under the curve between any two points.

Chapter

9

The Normal Curve

Key Terms

area under a curve
mu (μ)

normal curve
parameters

sigma (σ)
standard normal curve
z-scores

The **normal curve** is one of a number of possible models of probability distributions. Because it is widely used and is an important theoretical tool, it merits its own chapter in this book.

The normal curve is not a single curve; rather it is an infinite number of possible curves, all described by the same algebraic expression:

$$p(X) = \frac{1}{\sqrt{2\pi\sigma^2}}\ e^{\frac{-(X-\mu)^2}{2\sigma^2}}$$

The algebraic expression that describes the normal curve.

Upon viewing this expression for the first time, the initial reaction of most students is usually to panic. Don't. In general it is not necessary to "know" this formula to appreciate and use the normal curve. It is, however, useful to examine this expression for an understanding of how the normal curve operates.

First, some symbols in the expression are simply numbers. These symbols include 2, π, and e. The latter two are rational numbers that are very long, π equaling 3.1416... and e equaling 2.81.... As discussed in Chapter 3, "The Language of Algebra," it is possible to raise a "funny number," in this case e, to a "funny power."

The second set of symbols that are of some interest includes the symbol X, which is a variable corresponding to the score value. The height of the curve at any point is a function of X.

Thirdly, the final two symbols in the equation, μ and σ are called **parameters,** or values that, when set to particular numbers, define one of the infinite number of possible normal curves. The symbols μ and σ are Greek and are often written in English as **mu** and **sigma,** respectively. The concept of parameters is very important and considerable attention will be given them in the rest of this chapter.

9-1 A Family of Distributions

The normal curve is called a family of distributions. Each member of the family is determined by setting the parameters (μ and σ) of the model to a particular value (number). Because the μ parameter can take on any value, positive or negative, and the σ parameter can take on any positive value, the family of normal curves is quite large, consisting of an infinite number of members. This makes the normal curve a general-purpose model, able to describe a large number of naturally occurring phenomena, from test scores to the size of the stars.

9-1a Similarity of Members of the Family of Normal Curves

All the members of the family of normal curves, although different, have a number of properties in common. These properties include shape, symmetry, tails approaching but never touching the X-axis, and area under the curve:

- **Bell shape**—All members of the family of normal curves share the same bell shape, given the X-axis is scaled properly. Most of the area under the curve falls in the middle. The tails (ends) of the distribution approach the X-axis but never touch, with very little of the area under them.

- **Bilateral symmetry**—All members of the family of normal curves are bilaterally symmetric. That is, if any normal curve was drawn on a two-dimensional surface (a sheet of paper), cut out, and folded through the third dimension, the two sides would be exactly alike. Human beings are approximately bilaterally symmetrical, with right and left sides.

- **Tails never touch X-axis**—All members of the family of normal curves have tails that approach but never touch the X-axis. The implication of this property is that no matter how far you travel along the number line, in either the positive or negative direction, there will still be some area under any normal curve. Thus, in order to draw the entire normal curve you must have an infinitely long line. Because most of the area under any normal curve falls within a limited range of the number line, only that part of the line segment is drawn for a particular normal curve.

- **Area under curve totals 1.00**—All members of the family of normal curves have a total area of one (1.00) under the curve, as do all probability models or models of frequency distributions. This property, in addition to the property of symmetry, implies that the area in each half of the distribution is .50 or one half.

9-1b Area under a Curve

In statistics, the **area under a curve** represents theoretical relative frequency or probability. It permits the statistician to make decisions about the world based on a belief about what the world looks like rather than the limited information available in a sample of scores. For example, the statistician would advise the shoe store owner to purchase shoes to stock his or her shelves based on the area under a normal curve model of the world rather than the proportion of individuals in the sample who wore a particular size shoe. Area under a curve may seem like a strange notion to many introductory statistics students, so let's take a brief look at it.

Area is a familiar concept. For example, the area of a square is s^2, or side squared; the area of a rectangle is length times height; the area of a right triangle is one-half its base times its height; and the area of a circle is $\pi * r^2$ or πr^2. It is valuable to know these formulas if you are purchasing such things as carpeting, shingles, and so on.

Areas may be added or subtracted from one another to find some resultant area. For example, suppose you had an L-shaped room and wanted to purchase new carpet. You could find the area (and thus the amount of carpeting you would need to buy) by taking the total area of the larger rectangle and subtracting the area of the rectangle that was not needed, or you could divide the area into two rectangles, find the area of each, and add the areas together. Both procedures are illustrated in the following figure.

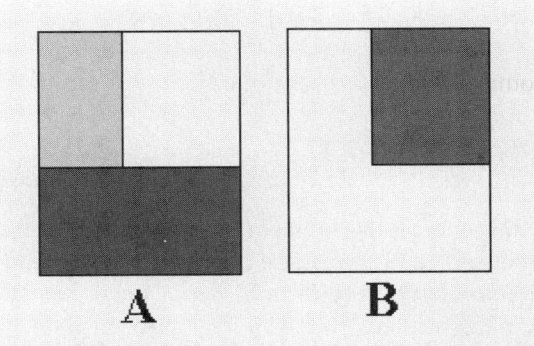

In A, the two shaded areas are added together; in B, the shaded area is subtracted from the whole area. Both ways have the same result: the area you need to know for purchasing your new carpet.

Finding the area under a curve poses a slightly different problem. In some cases there are formulas that directly give the area between any two points; finding these formulas are what integral calculus is all about. In other cases the areas must be approximated. Whether using integral calculus or approximating methods, the mathematician relies on the idea of adding together the areas of a number of rectangles.

Suppose a curve was divided into equally spaced intervals on the X-axis and rectangles were drawn corresponding to the height of the curve at each of the intervals. The rectangles may be drawn either smaller or larger than the curve, as the following two illustrations show:

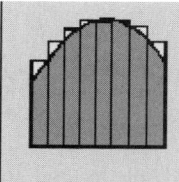

Approximating area under a curve by summing the larger rectangles.

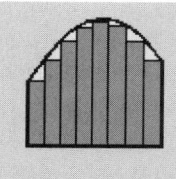

Approximating the area under a curve by summing the smaller rectangles.

In either case, if the areas of all the rectangles under the curve were added together, the sum of the areas would be an approximation of the total area under the curve. In the case of the smaller rectangles, the area would be too small; in the case of the larger, they would be too big. Taking the average would give a better approximation, but mathematical methods provide a better way.

A better estimate may be achieved by making the intervals on the X-axis smaller. Such an approximation, shown in the following illustration, would more closely measure the actual area under the curve.

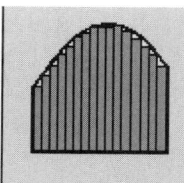

A better approximation using smaller intervals.

The actual area of the curve may be calculated by making the intervals infinitely small (no distance between the intervals) and then computing the area. If this last statement seems a bit bewildering, you share the bewilderment with millions of introductory calculus students. At this point the introductory statistics student must say, "I believe," and trust the mathematician or enroll in an introductory calculus course.

9-2 Drawing a Member of the Family of Normal Curves

The standard procedure for drawing a normal curve is to draw a bell-shaped curve and an X-axis. A tick is placed on the X-axis corresponding to the highest point (middle) of the curve. Three ticks are then placed to both the right and left of the middle point. These ticks are equally spaced and include all but a very small portion under the curve. The middle tick is labeled with the value of mu (μ); sequential ticks to the right are labeled by adding the value of sigma (σ). Ticks to the left are labeled by subtracting the value of sigma (σ) from μ for the three values. For example, if $\mu = 52$ and $\sigma = 12$, then the middle value would be labeled with 52; points to the right would have the values of 64 (52 + 12), 76, and 88; and points to the left would have the values 40, 28, and 16. Here's an illustration of this example:

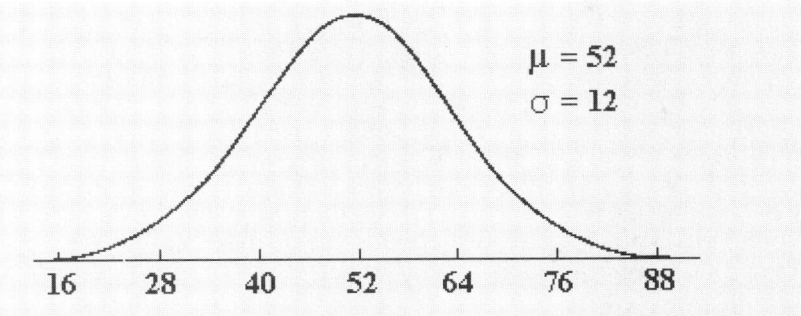

A normal curve with mu = 52 and sigma = 12.

9-3 Differences in Members of the Family of Normal Curves

Differences in members of the family of normal curves are a direct result of differences in values for parameters. The two parameters, μ and σ, each change the shape of the distribution in a different manner.

The first, μ, determines where the midpoint of the distribution falls. Changes in μ, without changes in σ, result in moving the distribution to the right or left, depending upon whether the new value of μ is larger or smaller than the previous value, but does not change the shape of the distribution. The following figure shows how a change in μ affects the normal curve.

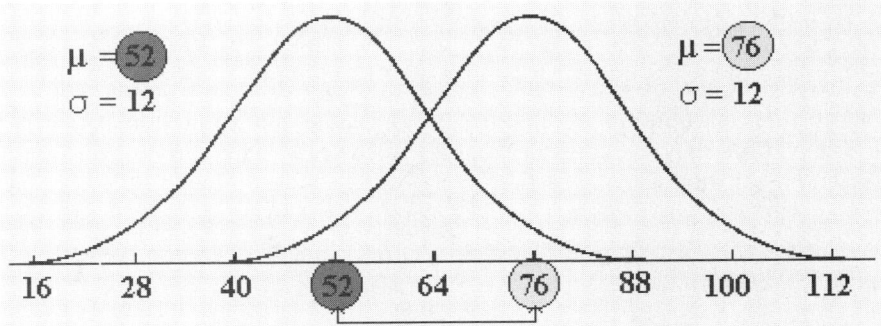

Comparing normal curves with a constant value of sigma = 12 and differing values of mu.

Changes in the value of σ, on the other hand, alter the shape of the distribution without affecting the midpoint, because σ affects the spread or the dispersion of scores. The larger the value of σ, the more dispersed the scores; the smaller the value, the less dispersed. Perhaps the easiest way to understand how σ affects the distribution is graphically. The following figure illustrates the effect of increasing the value of σ:

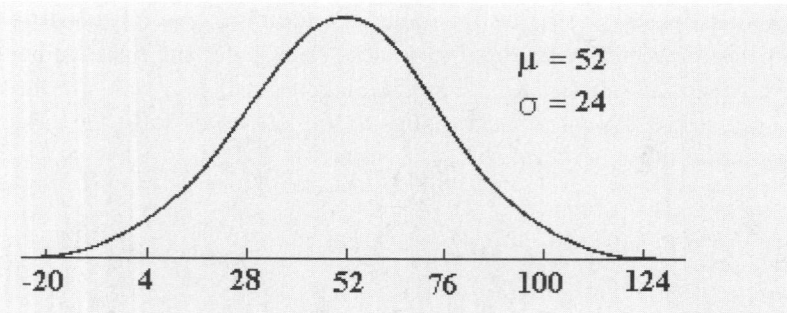

A normal curve with mu = 52 and sigma = 24.

Since this distribution was drawn according to the procedure described earlier, it appears similar to the previous normal curve, except for the values on the X-axis. This procedure effectively changes the scale and hides the real effect of changes in σ. Suppose the second distribution was drawn on a rubber sheet instead of a sheet of paper and stretched to twice its original length in order to make the two scales similar. Drawing the two distributions on the same scale results in the following graphic:

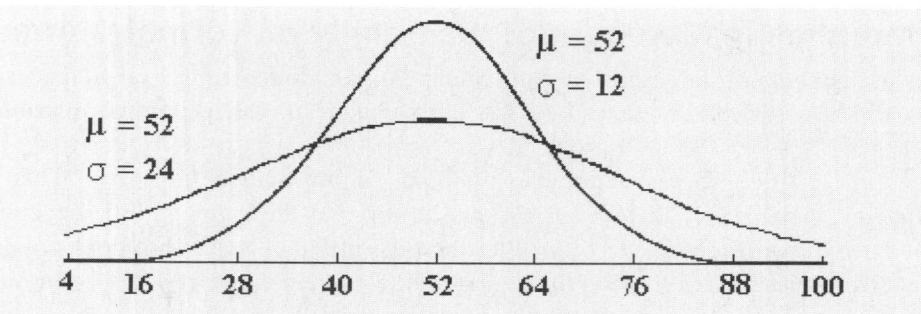

Comparing two normal curves with similar values for mu and different values for sigma.

Note that the shape of the second distribution has changed dramatically, being much flatter than the original distribution. It must not be as high as the original distribution because the total area under the curve must be constant, that is, 1.00. The second curve is still a normal curve; it is simply drawn on a different scale on the X-axis.

A different effect on the distribution may be observed if the size of σ is decreased. Here's a figure illustrating a new distribution depicted according to the standard procedure for drawing normal curves:

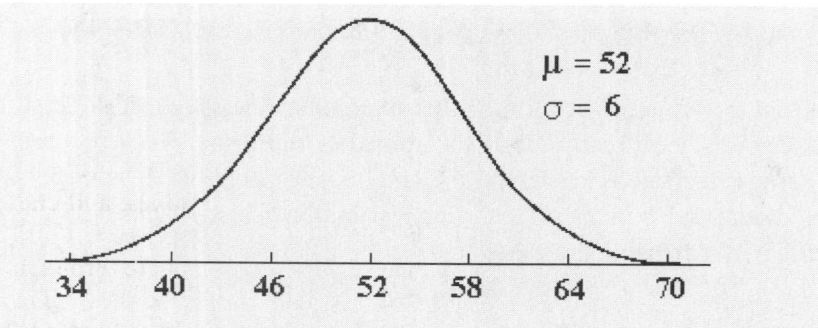

A normal curve with mu = 52 and sigma = 6.

Now both distributions are drawn on the same scale, as previously outlined, except in this case the sheet is stretched before the distribution is drawn and then released in order for the two distributions to be drawn on similar scales:

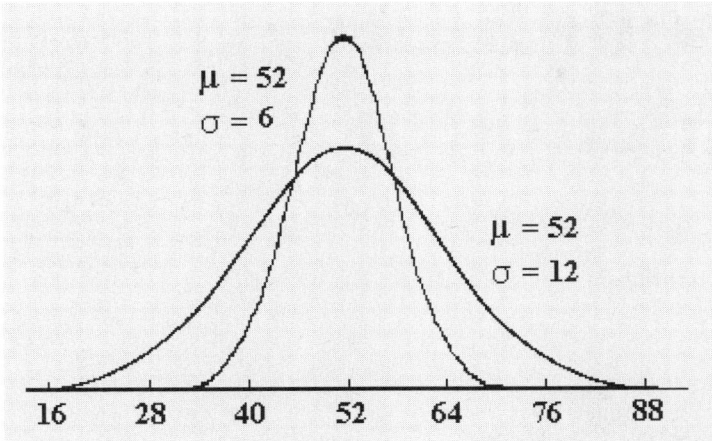

Comparing two normal curves with different values of sigma.

Note that the distribution is much higher in order to maintain the constant area of 1.00, and the scores are much more closely clustered around the value of μ, or the midpoint, than before.

9-4 Finding Area under Normal Curves

Suppose that when ordering shoes to restock the shelves in the store, you knew that female shoe sizes were normally distributed with $\mu = 7.0$ and $\sigma = 1.1$. Don't worry about where these values came from at this point—there will be plenty about that later. If the area under this distribution between 7.75 and 8.25 could be found, then you would know the proportion of size eight shoes to order. The values of 7.75 and 8.25 are the real limits of the interval of size eight shoes.

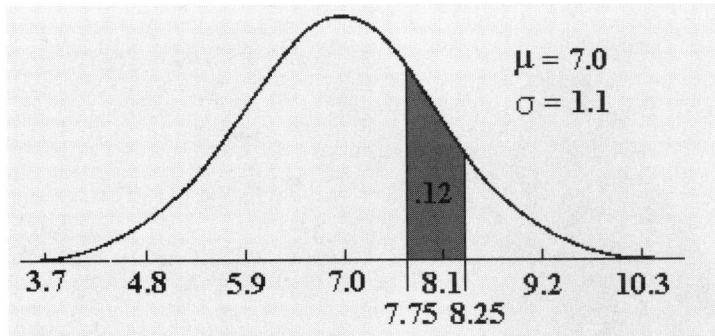

Area under a normal curve between the values of 7.75 and 8.25 where mu = 7.0 and sigma = 1.1.

Finding scores and areas on normal curves is easy using the Probability Calculator; simply select the normal distribution, enter values into the correct boxes, and click a button. The area or score will be entered in the correct box and a representation of the curve will be drawn in the display area. The following figure shows the initial screen of the Probability Calculator. (*Note:* The Probability Calculator can only be accessed by going to the online version of the text.)

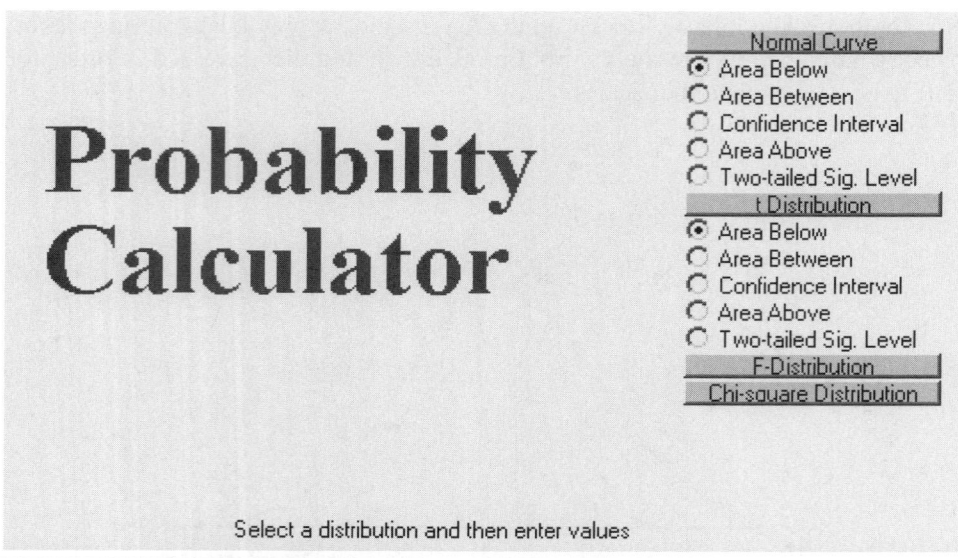

Select a distribution and then enter values

The initial screen of the Probability Calculator.

9-4a Area below a Score

To find the area below 7.75 on a normal curve with mu = 7.0 and sigma = 1.1, complete the following steps on the Probability Calculator.

1. Select Area Below under the Normal Curve button.

2. Click the Normal Curve button.

3. Enter values in the Mu, Sigma, and Score boxes.

4. Click the button with the arrow pointing to the right.

The steps and the result are presented in the following figure:

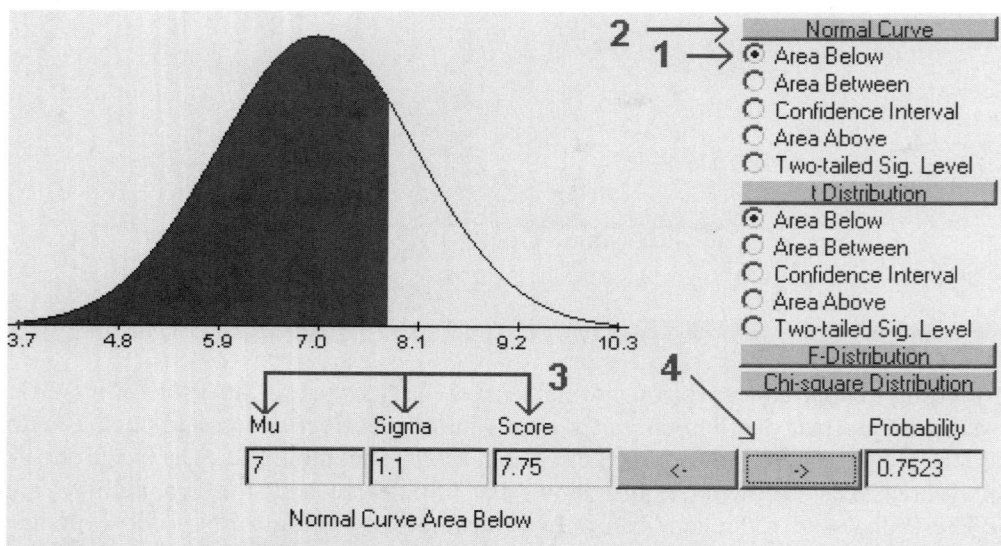

Finding area below a score in a given normal curve.

9-4b Area between Scores

To find the area between scores, use the Probability Calculator and follow these steps:

1. Select Area Between under the Normal Curve button.

2. Click the Normal Curve button.

3. Enter values in the Mu, Sigma, Low Score, and High Score boxes.

4. Click the button with the arrow pointing to the right.

The steps and the result are shown in the following figure.

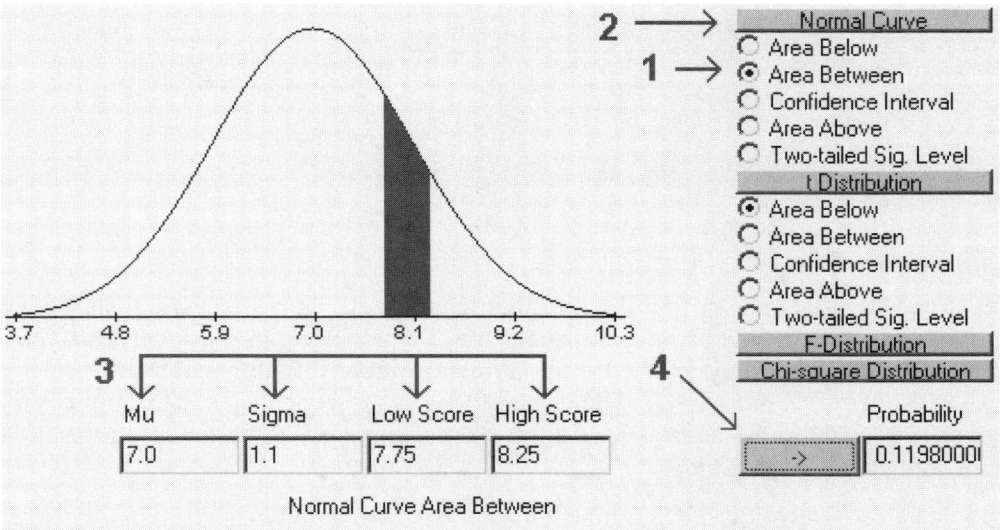

Finding the area under a normal curve between two scores.

9-4c Area above a Score

To find the area above a score, use the Probability Calculator and follow these steps:

1. Select Area Above under the Normal Curve button.

2. Click the Normal Curve button.

3. Enter values in the Mu, Sigma, and Score boxes.

4. Click the button with the arrow pointing to the right.

The following figure shows these steps and the result:

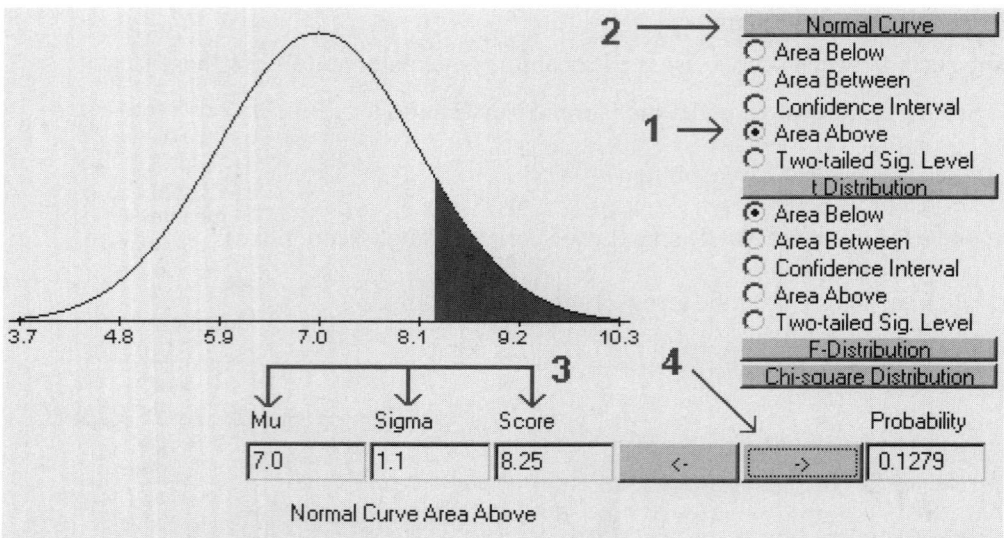

Finding area above a score under a normal curve.

9-4d Finding Probabilities in Both Tails

At times it will be necessary to find the area that falls in both the tail beyond a single score and the area in the tail of the mirror image of the score. For example, in a normal distribution with mu = 100, sigma = 3, and a score of 105.3, you might want to find the total area to the right of a value of 105.3 and to the left of 94.7. The value of 94.7 is the same distance from mu as the score of 105.3 (105.3 − 100 = 5.3 and 94.7 − 100 = -5.3). This area can be found on the Probability Calculator using the following steps.

1. Select "Two-tailed Sig. Level" under the Normal Curve button.

2. Click the Normal Curve button.

3. Enter values in the Mu, Sigma, and Score boxes.

4. Click the button with the arrow pointing to the right.

You can see the steps and the result in the following figure:

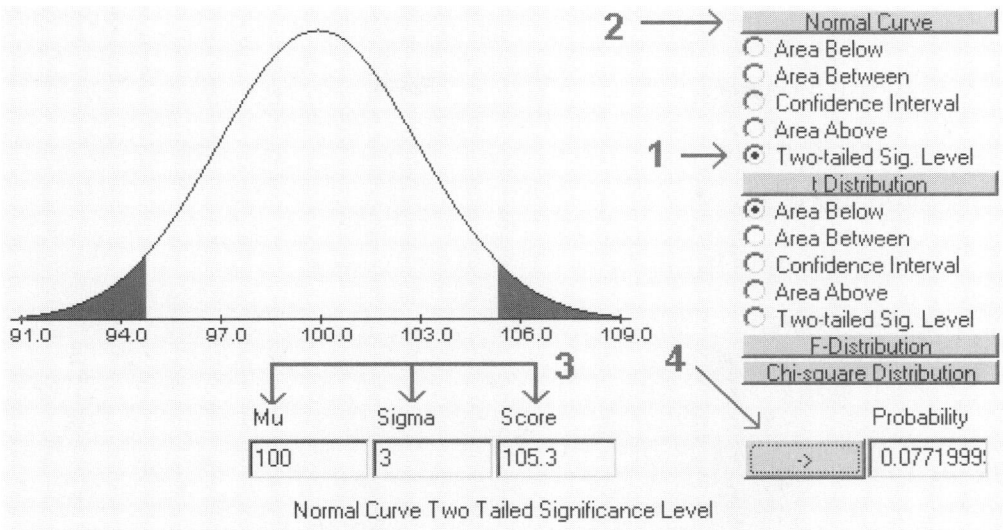

Finding two-tailed significance level under a normal curve.

If a value of 94.7 had been entered in the Score box in this illustration, the same result would have occurred.

9-5 Finding Scores from Area

In some applications of the normal curve, it will be necessary to find the scores that cut off some proportion or percentage of area of the normal distribution. One such application is finding two scores that cut off some symmetrical middle area of a given normal curve. The scores form a confidence interval around the middle point set by the value of mu. Confidence intervals are used to give a range of values rather than a single value to a given measure or estimate. Confidence intervals incorporate error or uncertainty into the information that is given to the client or reader. For example, if finding a ninety-five percent (.95 area) confidence interval resulted in a low score of 83.72 and a high score of 95.87, then ninety-five times out of one hundred the true score will fall between those two points.

Other applications, namely those of finding the percentile rank based on the normal curve, require the ability to find the score value that cuts off a given area of a normal curve. For example, a score value that cuts off 20 percent (.20 area) of a given normal curve would be said to have a percentile rank of 20.

The Probability Calculator has the ability to find both types of scores, given area.

It is important to note, however, that interpretation of scores obtained from the normal distributions in this manner are dependent upon the assumptions underlying their construction being reasonable. In order for a percentile rank based on the normal curve to be a valid estimate, the underlying distribution must be a normal distribution with the correct parameters of mu and sigma.

9-5a Finding Scores That Cut Off Middle Area

To use the Probability Calculator to find the scores that cut off some proportion or percentage of area of the normal distribution, it will be necessary to enter values for mu, sigma, and a probability. For example, suppose you wanted to know what two scores cut off the middle 75% of a normal distribution with $\mu = 123$ and $\sigma = 23$. The score values are used to find confidence intervals for various scores and statistics. In order to answer questions of this nature with the calculator, follow these steps:

1. Select Confidence Interval under the Normal Curve button.

2. Click the Normal Curve button.

3. Enter values in the Mu, Sigma, and Probability boxes.

4. Click the button with the arrow pointing to the left.

The steps and the result are shown in the following figure:

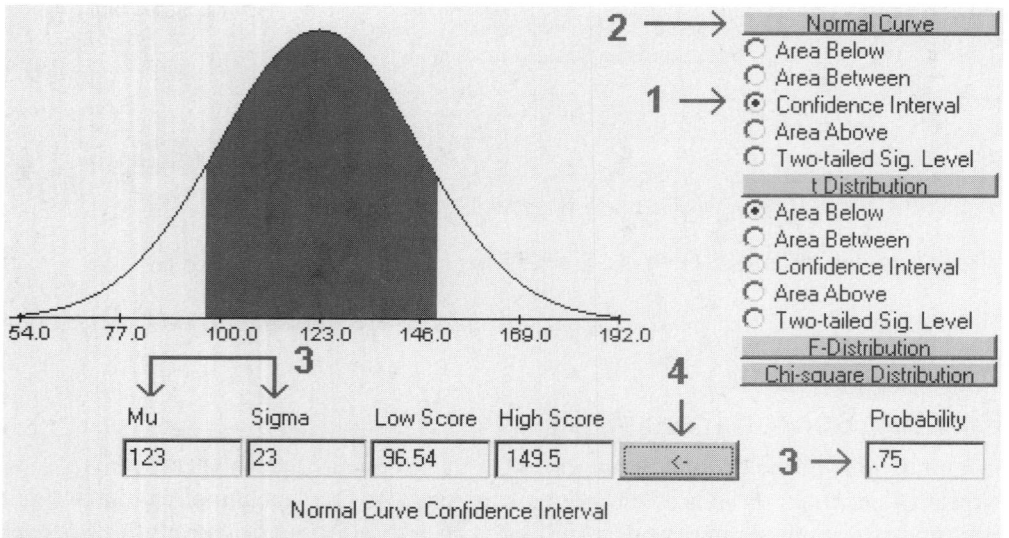

Finding scores that cut off a symmetrical middle area under a given normal curve.

In this case, the scores of 96.54 and 149.5 would be said to be form the seventy-five percent confidence interval in a normal distribution with mu = 123 and sigma = 23.

9-5b Finding Scores That Cut Off Bottom Area

In a similar manner, the score value that cuts off the bottom proportion of a given normal curve can be found using the Probability Calculator. For example, a score of 138.52 cuts off .75 of a normal curve with mu = 123 and sigma = 23. This area was found by using the following steps:

1. Select Area Below under the Normal Curve button.

2. Click the Normal Curve button.

3. Enter values in the Mu, Sigma, and Probability boxes.

4. Click the button with the arrow pointing to the left.

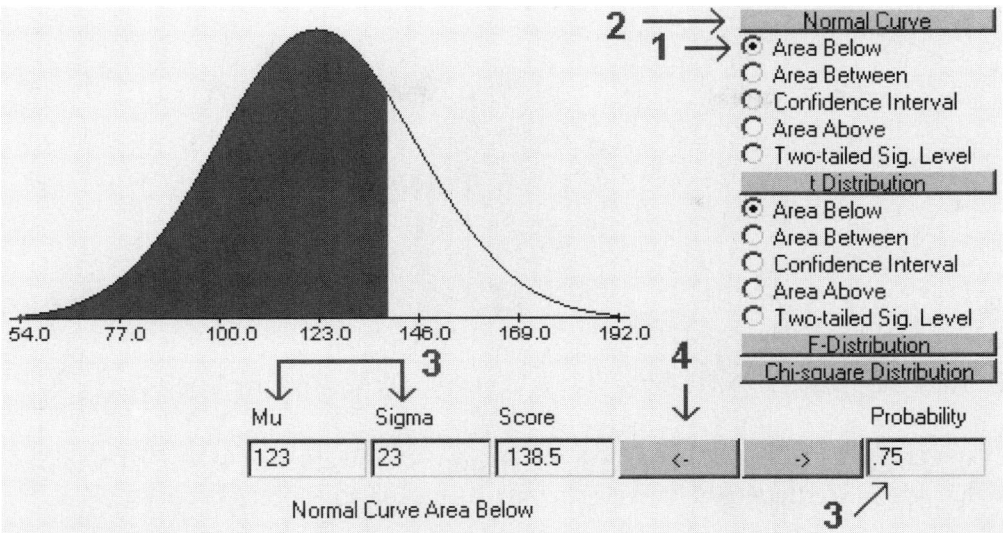

Finding the area below a given score on a given normal curve.

In this case, a score of 138.5 would have a percentile rank of seventy-five in a normal distribution with mu = 123 and sigma = 23.

9-6 The Standard Normal Curve

The **standard normal curve** is a member of the family of normal curves with $\mu = 0.0$ and $\sigma = 1.0$. The value of 0.0 was selected because the normal curve is symmetrical around μ and the number system is symmetrical around 0.0. The value of 1.0 for σ is simply a unit value. The X-axis on a standard normal curve is often relabeled and called **z-scores.**

There are three areas on a standard normal curve that all introductory statistics students should know:

1. **The area below 0.0 is .50 or 50%.** This is because the standard normal curve is symmetrical like all normal curves. This result generalizes to all normal curves in that the total area below (to the left of) the value of mu is .50 in any member of the family of normal curves.

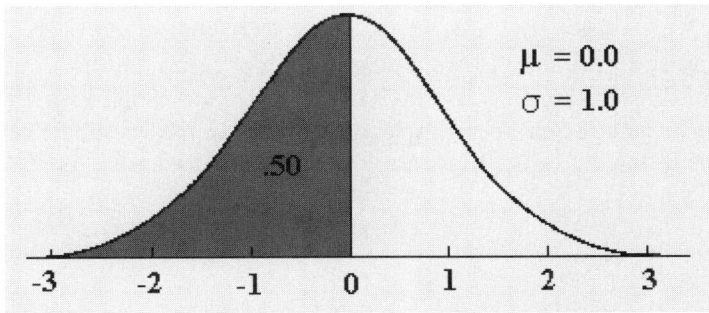

Half the area falls below the value of 0 on a standard normal curve.

2. **The area between z-scores of -1.00 and +1.00 is .68 or 68%.** The total area between plus and minus one sigma unit on any member of the family of normal curves is also .68.

The Normal Curve   **81**

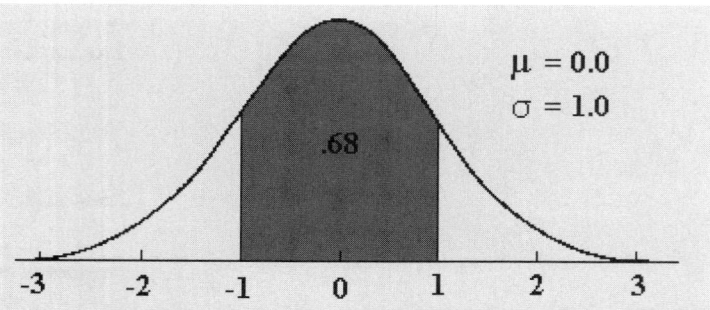

The area under a standard normal curve between plus and minus one sigma unit is .68.

3. **The area between z-scores of -2.00 and +2.00 is .95 or 95%.** This area (.95) also generalizes to plus and minus two sigma units on any normal curve.

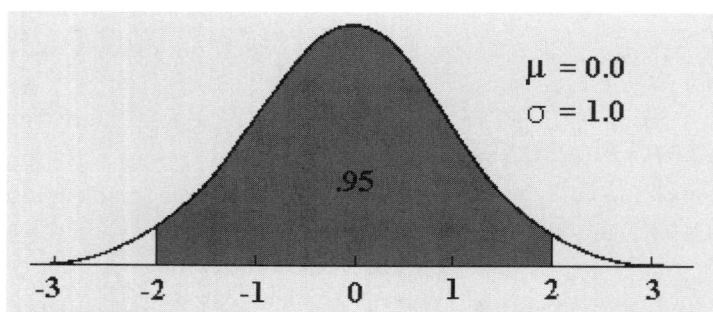

The area under a normal curve between plus and minus two sigma units is .95.

Knowing these areas allows computation of additional areas. For example, the area between a z-score of 0.0 and 1.0 may be found by taking 1/2 the area between z-scores of -1.0 and 1.0, because the distribution is symmetrical between those two points. The answer in this case is .34 or 34%. A similar logic and answer is found for the area between 0.0 and -1.0 because the standard normal distribution is symmetrical around the value of 0.0.

The area below a z-score of 1.0 may be computed by adding .34 and .50 to get .84. The area above a z-score of 1.0 may now be computed by subtracting the area just obtained from the total area under the distribution (1.00), giving a result of 1.00–.84 or .16 (16%).

The area between -2.0 and -1.0 requires additional computation. First, the area between 0.0 and -2.0 is 1/2 of .95 or .475. Because the .475 includes too much area, the area between 0.0 and -1.0 (.34) must be subtracted in order to obtain the desired result. The correct answer is .475–.34 or .135. The following figure illustrates the areas just discussed.

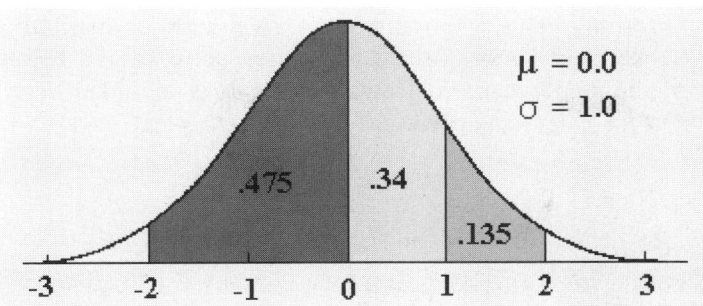

Area under various portions of a standard normal curve.

Using a similar kind of logic to find the area between z-scores of .5 and 1.0 will result in an incorrect answer because the curve is not symmetrical around .5. The correct answer must be something less than .17, because the desired area is on the smaller side of the total divided area. Because of this difficulty, the areas should be found using the Probability Calculator. Follow these steps to get the correct answer:

1. Select Area Between under the Normal Curve button.

2. Click the Normal Curve button.

3. Enter values in the Mu, Sigma, Low Score, and High Score boxes (use the values shown in the following figure).

4. Click the right-pointing arrow button.

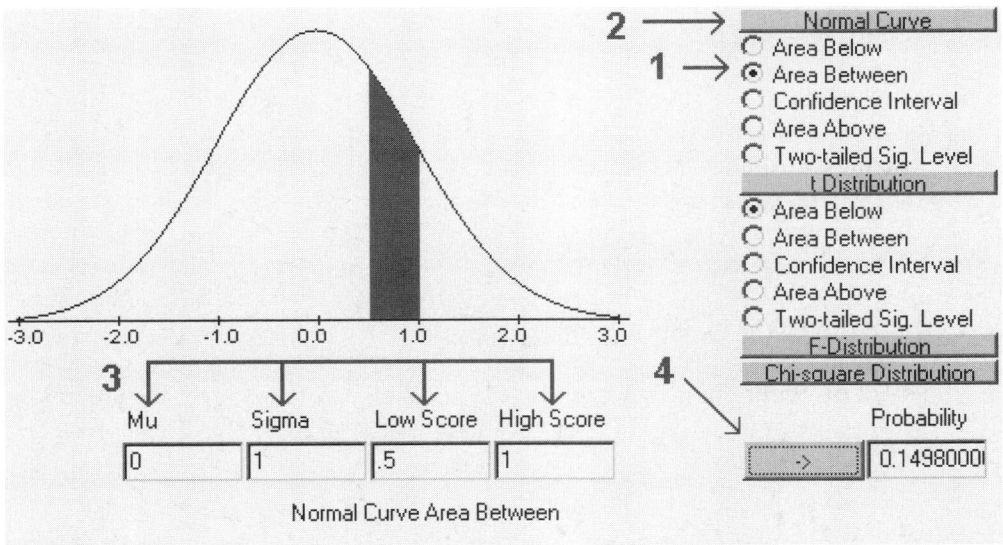

Finding the area under a standard normal curve between scores of .5 and 1.

The following formula is used to transform a given normal distribution into the standard normal distribution. It was much more useful when area between and below a score was only contained in tables of the standard normal distribution. It is included here for both historical reasons and because it will appear in a different form later in this text.

$$Z = \frac{X - \mu}{\sigma}$$

Formula for converting a score in a normal curve given mu and sigma to a standard normal curve.

Summary

The normal curve is an infinite number of possible probability models called a family of distributions. Each member of the family is described by setting the parameters (μ and σ) of the distribution to particular values. The members of the family are similar in that they share the same shape, are symmetrical, and have a total area underneath of 1.00. They differ in where the midpoint of the distribution falls, determined by μ, and in the variability of scores around the midpoint, determined by σ. The area between any two scores and the scores which cut off a given area on any given normal distribution can be easily found using the Probability Calculator provided in the online version of this text.

Chapter

10

Summation Notation

Key Terms

constant

subscripted variables

summation notation

summation sign (Σ)

It is necessary to enhance the language of algebra with an additional notational system in order to efficiently write some of the expressions that will be encountered in Chapter 11, "Statistics." The notational scheme provides a means of representing both a large number of variables and the summation of an algebraic expression.

10-1 Subscripted Variables

Suppose the following were scores made on the first homework assignment by five students in the class: 5, 7, 7, 6, and 8. These scores could be represented in the language of algebra by the symbols: V, W, X, Y, and Z. This method of representing a set of scores becomes awkward when the number of scores is greater than 26, so some other method of representation is necessary. The method of choice is called **subscripted variables,** written as X_i, where the X is the variable name, and the subscript (i) is a "dummy" or counter variable in that it may take on values from 1 to N, where N is the number of scores, to represent which score is being described. In the case of the example scores, then, $X_1 = 5$, $X_2 = 7$, $X_3 = 7$, $X_4 = 6$, and $X_5 = 8$.

If you wanted to represent the scores made on the second homework by these same students, the symbol Y_i could be used. The variable Y_1 would be the score made by the first student, Y_2 by the second student, and so on.

10-2 Summation Notation

Very often in statistics an algebraic expression of the form $X_1 + X_2 + X_3 + ... + X_N$ is used in a formula to compute a statistic. (The three dots in the expression mean that something is left out of the sequence and should be filled in when interpretation is done.) It is tedious to write an expression like this very often, so mathematicians have developed a shorthand notation to represent a sum of scores, called the **summation notation.**

The expression in front of the equals sign in what follows is summation notation; the expression that follows gives the meaning of the expression in "longhand" notation.

$$\sum_{i=1}^{N} X_i = X_1 + X_2 + ... + X_N$$

The summation sign defined.

The expression is read, "The sum of X sub i from i equals 1 to N." It means "add up all the numbers." In the example set of five numbers, where N = 5, the summation could be written:

$$\sum_{i=1}^{5} X_i = X_1 + X_2 + X_3 + X_4 + X_5 = 5 + 7 + 7 + 6 + 8 = 33$$

The i = 1 in the bottom of the summation notation tells where to begin the sequence of summation. If the expression were written with i = 3, the summation would start with the third number in the set. For example:

$$\sum_{i=3}^{N} X_i = X_3 + X_4 + ... + X_N$$

In the example set of numbers, this would produce the following result:

$$\sum_{i=3}^{N} X_i = X_3 + X_4 + X_5 = 7 + 6 + 8 = 21$$

The "N" in the upper part of the summation notation tells where to end the sequence of summation. If there were only three scores, then the summation and example would be:

$$\sum_{i=1}^{3} X_i = X_1 + X_2 + X_3 = 5 + 7 + 7 = 21$$

Sometimes, if the summation notation is used in an expression and the expression must be written a number of times, as in a proof, then a shorthand notation for the shorthand notation is employed. When the **summation sign (Σ)** is used without additional notation, then "i = 1" and "N" are assumed. For example:

$$\sum X = \sum_{i=1}^{N} X_i = X_1 + X_2 + \dots + X_N$$

10-3 Summation of Algebraic Expressions

10-3a The General Rule

The summation notation may be used not only with single variables, but also with algebraic expressions containing more than one variable. When these expressions are encountered, considerable attention must be paid to where the parentheses are located. If the parentheses are located after the summation sign, then the general rule is:

Do the algebraic operation and then sum.

For example, suppose that a small portion of a teacher's grade book appears as follows, where X is the score for first homework assignment and Y is the score for the second:

Example X and Y Data	
X	Y
5	6
7	7
7	8
6	7
8	8

The sum of the product of the two variables could be written:

$$\sum_{i-1}^{N} (X_i * Y_i) = (X_1 * Y_1) + (X_2 * Y_2) + \dots + (X_N * Y_N)$$

The sum may be most easily computed by creating a third column on the data table, as shown here:

Sum of X Times Y		
X	Y	X*Y
5	6	30
7	7	49
7	8	56
6	7	42
8	8	64
33	36	241

And the summation expression would be completed as:

$$\sum_{i-1}^{5} (X_i * Y_i) = 30 + 49 + 56 + 42 + 64 = 241$$

Note that a change in the position of the parentheses dramatically changes the results:

$$(\sum_{i=1}^{N} X_i) * (\sum_{i=1}^{N} Y_i) = 33 * 36 = 1188$$

A similar kind of differentiation is made between ΣX^2 and $(\Sigma X)^2$. In the former the sum would be 223, while the latter would be 332 or 1089.

10-3b Exceptions to the General Rule

Three exceptions to the general rule provide the foundation for some simplification and statistical properties to be discussed later. These exceptions are

1. When the expression being summed contains a "+" or "−" at the highest level, then the summation sign may be taken inside the parentheses. The rule may be more concisely written:

$$\sum_{i-1}^{N} (X_i + Y_i) = \sum_{i=1}^{N} X_i + \sum_{i=1}^{N} Y_i$$

Computing both sides from a table with example data yields:

Sum of X + Y and X – Y			
X	Y	X + Y	X – Y
5	6	11	-1
7	7	14	0
7	8	15	-1
6	7	13	-1
8	8	16	0
33	36	69	-3

Note that the sum of the X + Y column is equal to the sum of the X column plus the sum of Y column. Similar results hold for the X – Y column.

2. **The sum of a constant times a variable is equal to the constant times the sum of the variable.** A **constant** is a value that does not change with the different values for the counter variable (I), such as numbers. If every score is multiplied by the same number and then summed, it would be equal to the sum of the original scores times the constant. Constants are usually identified in the statement of a problem, often represented by the letters c or k. If c is a constant, then, as before, this exception to the rule may be written in algebraic form:

$$\sum_{i=1}^{N} (c * X_i) = c * \sum_{i=1}^{N} X_i$$

For example, suppose that the constant was equal to 5. Using the example data produces the following result:

X and a Constant Times X	
X	5 * X
5	25
7	35
7	35
6	30
8	40
33	165

Note that c * 33 = 165, the same as the sum of the second column.

3. **The sum of a constant is equal to N times the constant.** If no subscripted variables (non-constants) are included on the right of a summation sign, then the number of scores is multiplied times the constant appearing after the summation. Writing this exception to the rule in algebraic notation:

$$\sum_{i=1}^{N} c = N * c$$

For example, if c = 8 and N = 5 then:

$$\sum_{i=1}^{N} c = \sum_{i=1}^{5} 8 = 8 + 8 + 8 + 8 + 8 = 5 * 8 = N * c$$

10-3c Solving Algebraic Expressions with Summation Notation

When algebraic expressions include summation notation, simplification can be performed if few rules are remembered.

1. The expression to the right of the summation sign may be simplified using any of the algebraic rewriting rules.

2. The entire expression including the summation sign may be treated as a phrase in the language.

3. The summation sign is *not* a variable, and may not be treated as one (cancelled for example.)

4. The three exceptions to the general rule may be used whenever applicable.

Two examples follow with X and Y as variables and c, k, and N as constants:

$$\frac{\sum (X+Y) + \sum X - \sum Y}{\sum X} =$$

$$\frac{\sum X + \sum Y + \sum X - \sum Y}{\sum X} =$$

$$\frac{2 * \sum X}{\sum X} = 2$$

Example summation simplification.

$$\frac{\sum (X^2 + 2XY + Y^2) - \sum (X^2 - 2XY + Y^2)}{8 * \sum XY} =$$

$$\frac{\sum X^2 + \sum 2XY + \sum Y^2 - \sum x^2 + \sum 2XY - \sum Y^2}{8 * \sum XY} =$$

$$\frac{\sum 2XY + \sum 2XY}{8 * \sum XY} =$$

$$\frac{2 * \sum 2XY}{8 * \sum XY} =$$

$$\frac{2 * 2 \sum XY}{8 * \sum XY} =$$

$$\frac{4 * \sum XY}{8 * \sum XY} =$$

$$\frac{1}{2}$$

Example summation sign simplification.

Summary

Use of additional shorthand notation, called summation notation, within the language of algebra allows statistical expressions that require the representation of a large number of variables to be written more efficiently. The notational scheme provides a means of representing the summation of an algebraic expression.

In this chapter you learned about using subscripted variables, written as X_i, where the X is the variable name and the subscript (i) is a counter variable that can take on values from 1 to N. Subscripted variables are very useful in representing sets of data. You were introduced to summation notation, a shorthand notation developed by mathematicians to represent a sum of scores, and saw several examples of the notation.

There was also a discussion of using summation notation with algebraic expressions, and you learned the importance of the location of parentheses in these expressions, as well as how to simplify expressions containing the summation notation.

With this information in hand, you're ready for the next chapter, "Statistics."

Chapter

11

Statistics

Key Terms

bimodal
central tendency
mean
median
mode

negatively skewed distribution
positively skewed distribution
range
skewed distribution
standard deviation

symmetrical distribution
variability
variance

A statistic is an algebraic expression combining scores into a single number. Statistics serve two functions: they estimate parameters in population models and they describe the data. The statistics discussed in this chapter will be used both as estimates of μ and σ and as measures of central tendency and variability. There are a large number of possible statistics, but some are more useful than others.

11-1 Measures of Central Tendency

Central tendency is a typical or representative score. If the mayor is asked to provide a single value which best describes the income level of the city, he or she would answer with a measure of central tendency. There are many possible measures of central tendency; the three major ones—the mode, median, and mean—will be discussed in this chapter.

11-1a The Mode

The **mode,** symbolized by M_o, is the most frequently occurring score value. If the scores for a given sample distribution are

Example Data													
32	32	35	36	37	38	38	39	39	39	40	40	42	45

then the mode would be 39 because a score of 39 occurs 3 times, more than any other score. The mode may be seen on a frequency distribution as the score value which corresponds to the highest point. For example, the following is a frequency polygon of the example data:

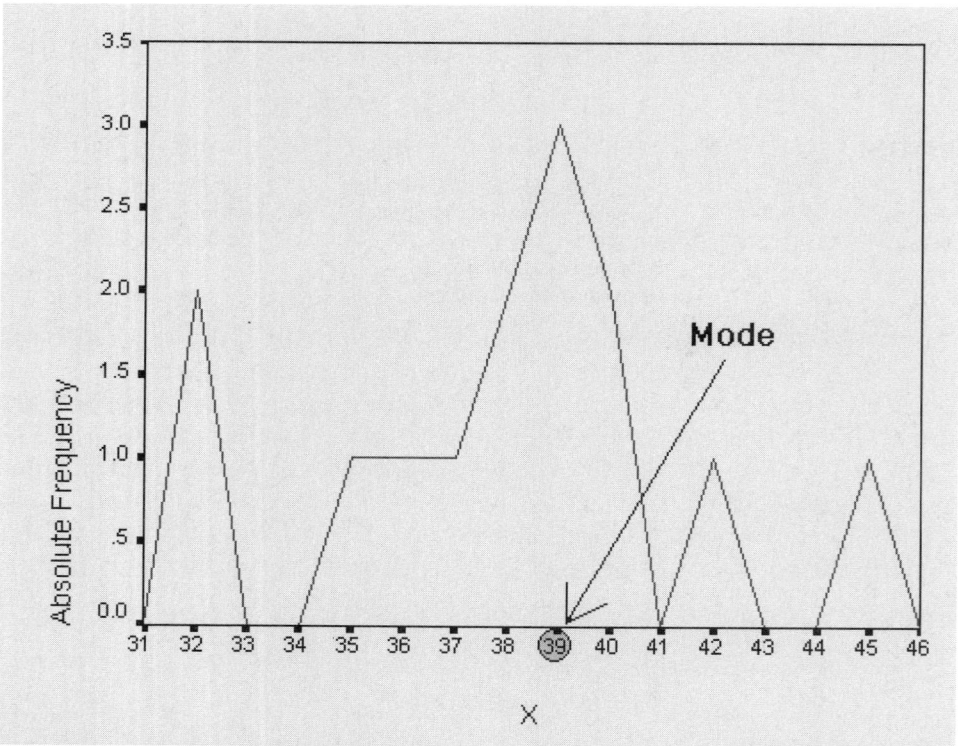

Locating the mode in a frequency polygon.

A distribution may have more than one mode if the two most frequently occurring scores occur the same number of times. For example, if the earlier score distribution were modified as follows

Example Bimodal Data													
32	32	32	36	37	38	38	39	39	39	40	40	42	45

then there would be two modes, 32 and 39. Such distributions are called **bimodal.** The frequency polygon of this bimodal distribution follows.

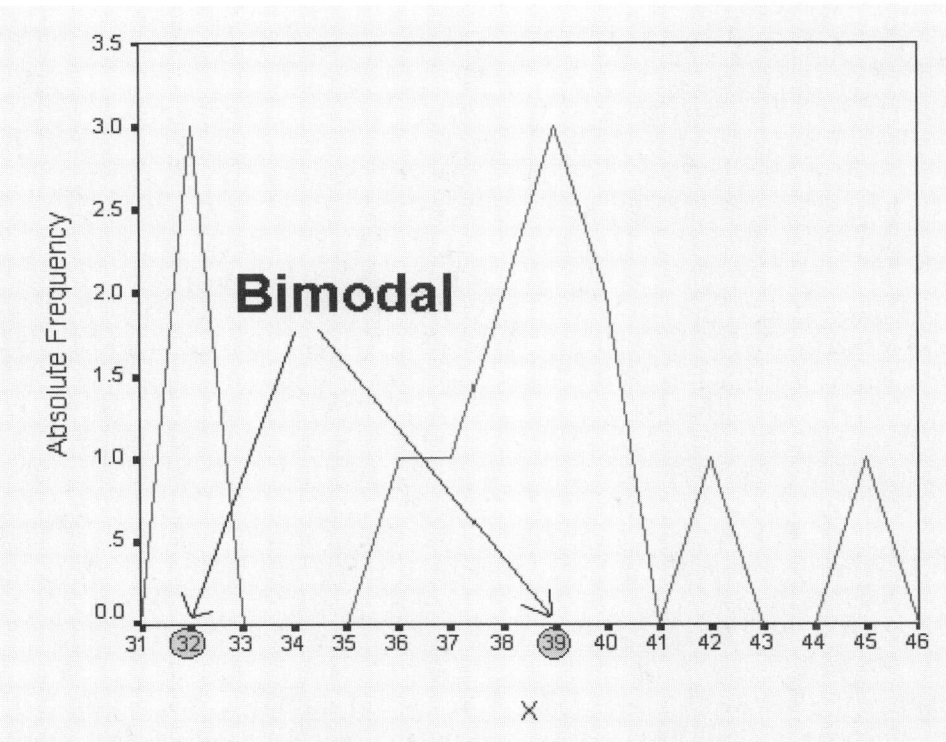

A bimodal frequency polygon.

In an extreme case there may be no unique mode, as in the case of a rectangular distribution.

The mode is not sensitive to extreme scores. Suppose the original distribution was modified by changing the last number, 45, to 55 as follows:

Example Data with Extreme Score													
32	32	35	36	37	38	38	39	39	39	40	40	42	55

The mode would still be 39, and the frequency polygon would look like this:

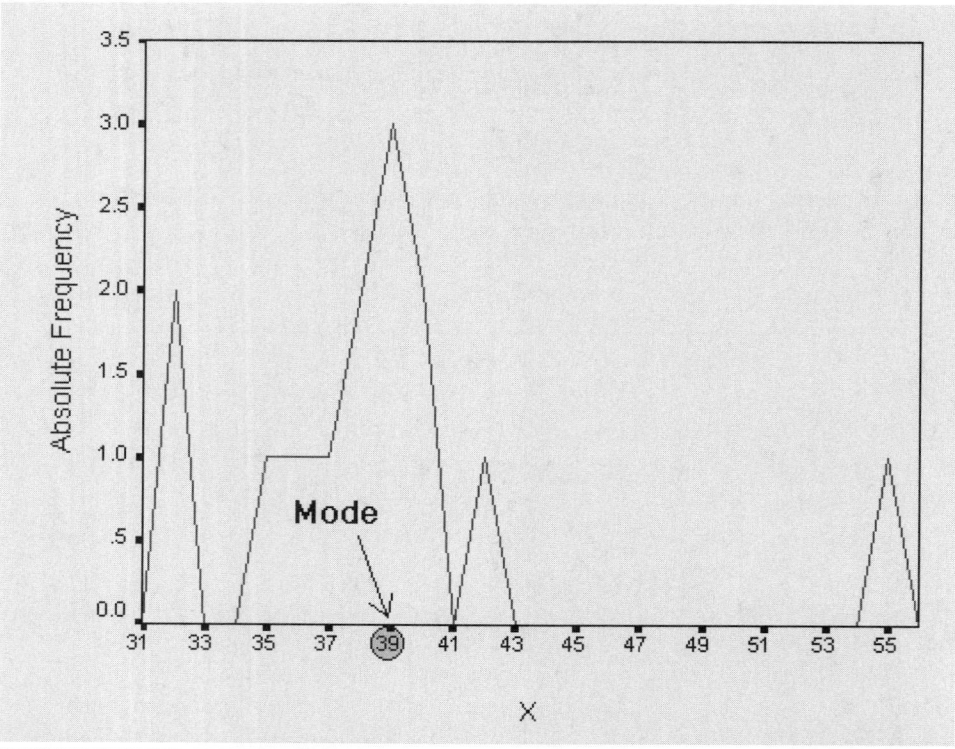

The mode is not sensitive to extreme scores.

In any case, the mode is a quick-and-dirty measure of central tendency—quick because it is easily and quickly computed, and dirty because it is not very useful (does not give much information about the distribution).

11-1b The Median

The **median,** symbolized by M_d, is the score value that cuts the distribution in half, such that half the scores fall above the median and half fall below it. Computation of the median is relatively straightforward. The first step is to rank order the scores from lowest to highest. The procedure branches at the next step: one way if there are an odd number of scores in the sample distribution, and another if there are an even number of scores.

If there is an odd number of scores, as in this distribution,

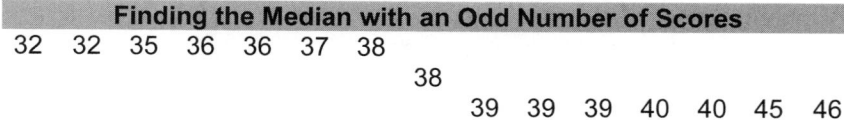

Finding the Median with an Odd Number of Scores

32 32 35 36 36 37 38

38

39 39 39 40 40 45 46

then the median is simply the middle number. In this case the median would be the number 38, because there are 15 scores all together with 7 scores smaller and 7 larger.

If there is an even number of scores, as in this distribution,

Finding the Median with an Even Number of Scores														
32	35	36	36	37	38									
						38	39							
								39	39	40	40	42	45	

then the median is the midpoint between the two middle scores: in this case the value 38.5. It was found by adding the two middle scores together and dividing by two: (38 + 39) / 2 = 38.5. If the two middle scores are the same value, then the median is that value.

In this system, no account is paid to whether there is a duplication of scores around the median. In some systems a slight correction is performed to correct for grouped data, but since the correction is slight and the data is generally not grouped for computation in calculators or computers, it is not presented here.

The median, like the mode, is not affected by extreme scores, as the following distribution of scores indicates:

Finding the Median with an Extreme Score														
32	35	36	36	37	38									
						38	39							
								39	39	40	40	42	55	

The median is still the value of 38.5. The median is not as quick-and-dirty as the mode, but generally it is not the preferred measure of central tendency.

11-1c The Mean

The **mean,** symbolized by $\overline{X}$, is the sum of the scores divided by the number of scores. The following formula both defines and describes the procedure for finding the mean:

$$\overline{X} = \frac{\sum_{i=1}^{N} X_i}{N}$$

The mean is equal to the sum of the scores divided by the number of scores.

where ΣX is the sum of the scores and N is the number of scores. Application of this formula to the following data

Example Data to Find the Mean												
32	35	36	37	38	38	39	39	39	40	40	42	45

yields the following results:

$$\overline{X} = \frac{\sum X}{N} = \frac{500}{13} = 38.46$$

The mean.

Use of means as a way of describing a set of scores is fairly common; batting average, bowling average, grade point average, and average points scored per game are all means. Note the use of the word "average" in all of those examples. In most cases when the term "average" is used, it refers to the mean, although not necessarily. When a politician uses the term "average income" for example, he or she may be referring to the mean, median, or mode.

The mean is sensitive to extreme scores. For example, the mean of the following data is 39.0, somewhat larger than the preceding example.

Finding the Mean with an Extreme Score
32   35   36   37   38   38   39   39   39   40   40   42   55

In most cases the mean is the preferred measure of central tendency, both as a description of the data and as an estimate of the parameter. In order for the mean to be meaningful, however, the acceptance of the interval property of measurement is necessary. When this property is obviously violated, it is inappropriate and misleading to compute a mean. Such is the case, for example, when the data are clearly nominal-categorical. An example would be political party preference where 1 = Republican, 2 = Democrat, and 3 = Independent. The special case of dichotomous nominal categorical variables allows meaningful interpretation of means. For example, if only two levels of political party preference was allowed, 1 = Republican and 2 = Democrat, then the mean of this variable could be interpreted. In such cases it is preferred to code one level of the variable with a 0 and the other level with a 1 such that the mean is the proportion of the second level in the sample. For example, if gender was coded with 0 = Male and 1 = Female, then the mean of this variable would be the proportion of females in the sample.

11-1d Exercise: Kiwi Bird Problem

A Kiwi bird.

As is commonly known, Kiwi birds are native to New Zealand. They are born exactly one foot tall and grow in one-foot intervals. That is, one moment they are one foot tall and the next they are two feet tall. They are also very rare. An investigator goes to New Zealand and finds four birds. The mean height of the four birds is 4, the median is 3, and the mode is 2. What are the heights of the four birds?

Hint: Examine the constraints of the mode first, the median second, and the mean last.

11-1e Skewed Distributions and Measures of Central Tendency

Skewness refers to the asymmetry of the distribution, such that a symmetrical distribution exhibits no skewness. In a **symmetrical distribution,** the mean, median, and mode all fall at the same point, as in the following distribution:

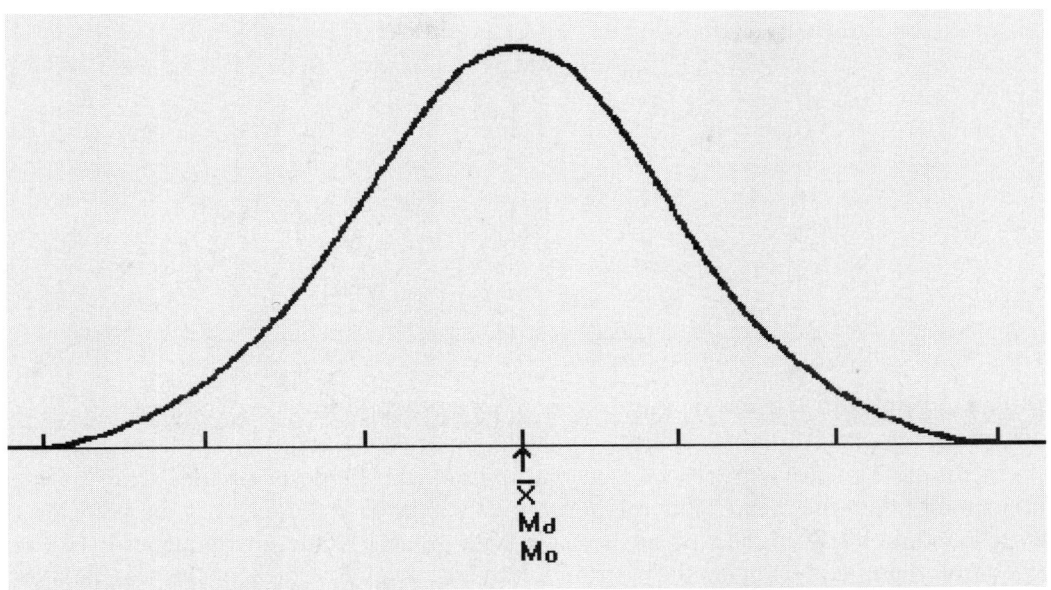

The mean, median, and mode have the same value in a normal curve.

An exception to this is the case of a bimodal symmetrical distribution, where the mean and the median fall at the same point, while the two modes correspond to the two highest points of the distribution. Here's an example:

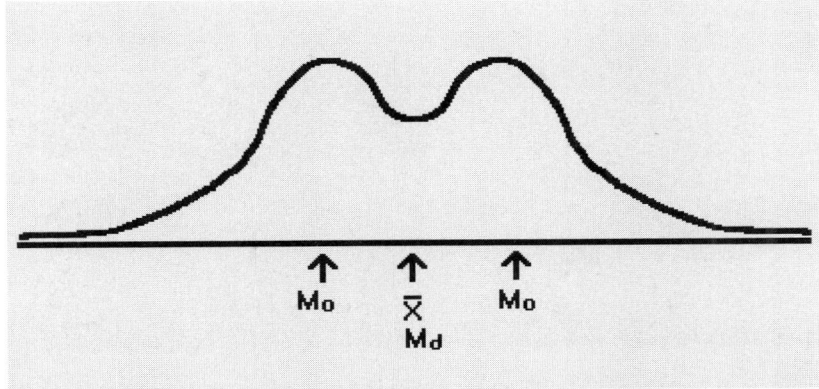

The mean, median, and modes in a bimodal symmetrical distribution.

A **positively skewed distribution** is asymmetrical and points in the positive direction. If a test was very difficult and almost everyone in the class did very poorly on it, the resulting distribution would most likely be positively skewed, as the following distribution shows:

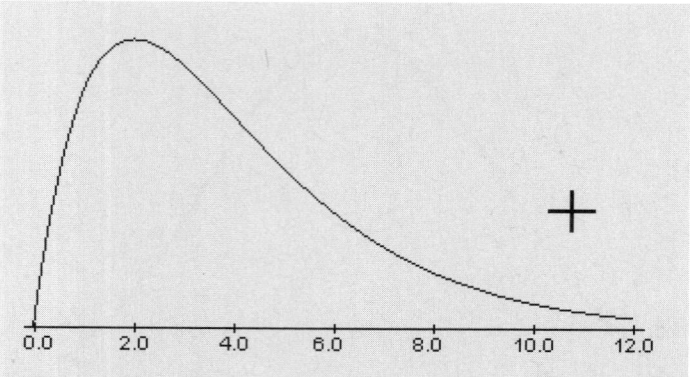

A positively skewed distribution.

In the case of a positively skewed distribution, the mode is smaller than the median, which is smaller than the mean. This relationship exists because the mode is the point on the x-axis corresponding to the highest point(the score with greatest value, or frequency). The median is the point on the x-axis that cuts the distribution in half, such that 50% of the area falls on each side, as shown in the following distribution:

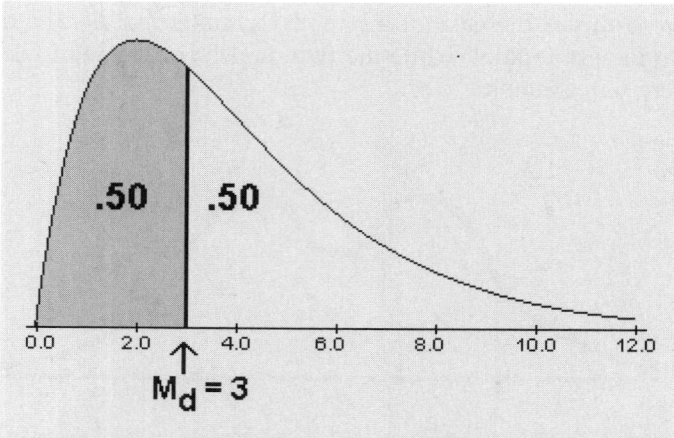

The median in a positively skewed distribution.

The mean is the balance point of the distribution. Because points further away from the balance point change the center of balance, the mean is pulled in the direction the distribution is skewed. For example, if the distribution is positively skewed, the mean would be pulled in the direction of the skewness, or be pulled toward larger numbers, as you can see in the distribution that follows.

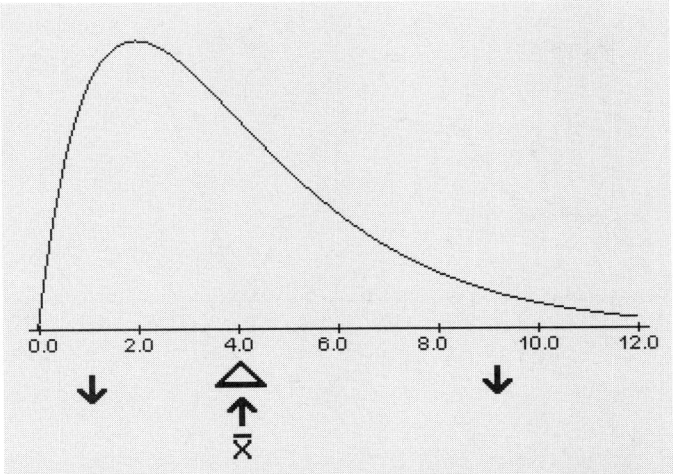

The mean as a balance point.

One way to remember the order of the mean, median, and mode in a **skewed distribution** is to remember that the mean is pulled in the direction of the extreme scores. In a positively skewed distribution, the extreme scores are larger, thus the mean is larger than the median, which you can see in the following distribution.

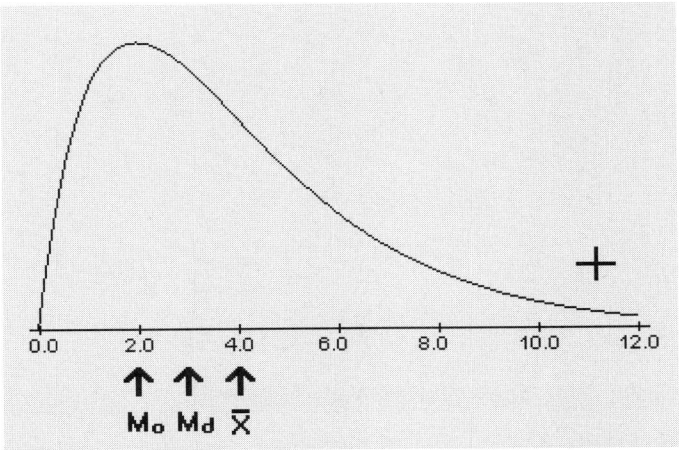

The mode, median, and mean in a positively skewed distribution.

A **negatively skewed distribution** is asymmetrical and points in the negative direction, such as would result with a very easy test. On an easy test, almost all students would perform well and only a few would do poorly. Here's an example of such a distribution:

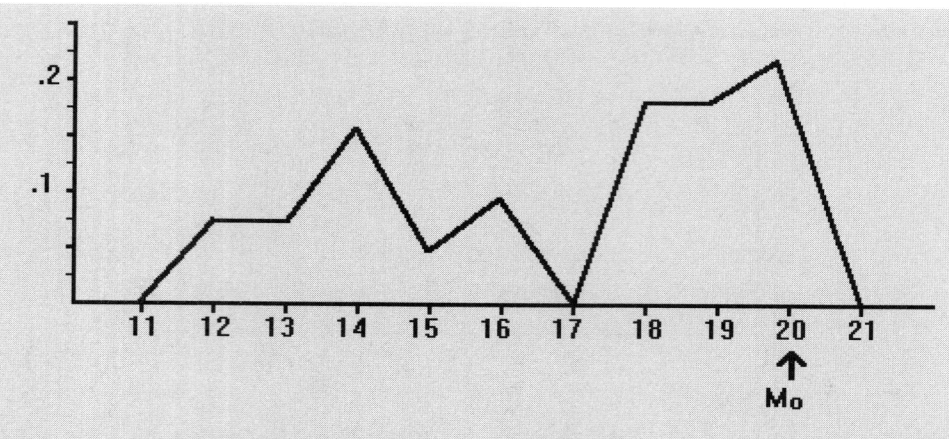

The mode in a negatively skewed distribution.

The order of the measures of central tendency in a negatively skewed distribution would be the opposite of the positively skewed distribution, with the mean being smaller than the median, which is smaller than the mode, as indicated in the following distribution:

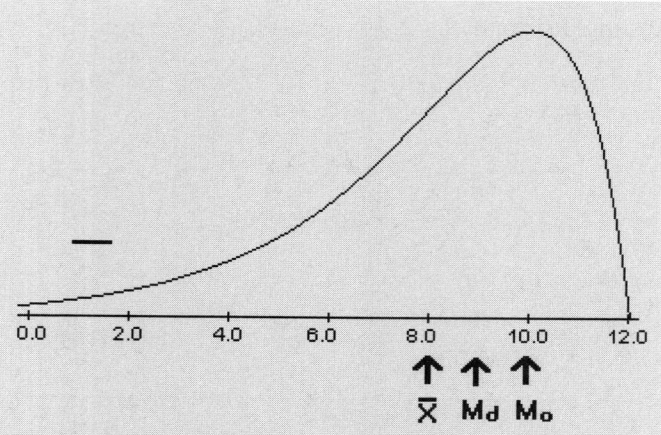

The mean, median, and mode in a negatively skewed distribution.

11-2 Measures of Variability

Variability refers to the spread or dispersion of scores. A distribution of scores is said to be highly variable if the scores differ widely from one another. Let's take a look at three statistics that measure variability: the range, the variance, and the standard deviation. The latter two are very closely related and will be discussed in the same section.

11-2a The Range

The **range** is the largest score minus the smallest score. It is a quick and dirty measure of variability, although when a test is given back to students they very often want to know the range of scores. Because the range is greatly affected by extreme scores, it may give a distorted picture of the scores. The following two distributions have the same range, 13, yet appear to differ greatly in the amount of variability.

Two Distributions with the Same Range but Different Variability											
Distribution 1	32	35	36	36	37	38	40	42	42	43	43	45
Distribution 2	32	32	33	33	33	34	34	34	34	34	35	45

For this reason, among others, the range is not the most important measure of variability.

11-2b The Variance and the Standard Deviation

The **variance**, symbolized by "s^2," is a measure of variability. The **standard deviation,** symbolized by "s," is the positive square root of the variance. It is easier to define the variance with an algebraic expression than words, thus the following formula:

$$s^2 = \frac{\sum_{i-1}^{N}(X_i - \bar{X})^2}{N-1}$$

Note that the variance could almost be the average squared deviation around the mean if the expression were divided by N rather than N–1. It is divided by N–1, called the degrees of freedom (df), for theoretical reasons. If the mean is known, as it must be to compute the numerator of the expression, then only N–1 scores are free to vary. That is, if the mean and N–1 scores are known, it is possible to figure out the Nth score. You only need to recall the Kiwi bird problem to convince yourself that this is true. When the heights of three of the four birds were known, it was possible to find the height of the fourth bird if the mean was also known. Similar logic works no matter what the size of N.

The formula for the variance is a definitional formula—it defines what the variance means. The variance may be computed from this formula, but in practice this is rarely done. It is done here to better describe what the formula means. The computation is performed in a number of steps, which are presented here:

Steps in Computing the Standard Deviation	
Step One	Find the mean of the scores.
Step Two	Subtract the mean from every score.
Step Three	Square the results of Step Two.
Step Four	Sum the results of Step Three.
Step Five	Divide the results of Step Four by N – 1.
Step Six	Take the square root of Step Five.

The result at step five is the sample variance; at step six, the sample standard deviation.

For the next example there are five scores: 8, 8, 9, 12, and 13. Refer to the following data table to complete the exercise.

Finding the Sum of Squared Deviations from the Mean		
X	X– $\bar{X}$	(X– $\bar{X}$)^2
8	-2	4
8	-2	4
9	-1	1
12	2	4
13	3	9
50	0	22

Following the six steps outlined previously, let's find the standard deviation of the sample scores:

1. Find the mean of the scores by summing the first column (50) and dividing by the number of scores (5): The mean is 10.

2. Subtract the mean (10) from each score. The results are shown in the second column of the preceding table. For example, the first score is 8; subtract the mean (10), and the result is –2.

3. Square the results in the second column and place those results in the third column. For example, the first entry, -2, squared is 4.

4. Sum the third column to find the sum of squared deviations from the mean, sometimes referred to as the sum of squares. This value is the numerator for the formula for the sample variance.

5. Divide the result of step 4 by the number of scores minus 1 to find the sample variance. In this case the sum of squares (22) is divided by 4—the number of scores (5) minus 1—to produce a result of 5.5.

6. Take the square root of the previous result (5.5) to obtain the standard deviation: 2.345.

Here are the same steps, shown in table form:

Steps in Computing the Standard Deviation		
Step One	Find the mean of the scores.	$\overline{X} = 50 / 5 = 10$
Step Two	Subtract the mean from every score.	The second column in the data table.
Step Three	Square the results of Step Two.	The third column in the data table.
Step Four	Sum the results of Step Three.	22
Step Five	Divide the results of Step Four by N – 1.	$s^2 = 22/4 = 5.5$
Step Six	Take the square root of Step Five.	$s = 2.345$

Note that in the data table, the sum of the second column is zero. This must be the case if the calculations are performed correctly up to that point.

The standard deviation measures variability in units of measurement, while the variance does so in units of measurement squared. For example, if you measured height in inches, then the standard deviation would be in inches, while the variance would be in inches squared. For this reason, the standard deviation is usually the preferred measure when describing the variability of distributions. The variance, however, has some unique properties that make it very useful in theoretical constructs. These properties will be discussed later in this text.

11-2c Calculating Statistics with a Statistical Calculator

Calculations may be checked by using the statistical functions of a statistical calculator. This is the way the variance and standard deviation are usually computed in practice. The calculator has the definitional formula for the variance programmed internally, and all that is necessary for its use are the following steps:

Steps in Computing the Standard Deviation Using a Statistical Calculator
Step One   Select the statistics mode.
Step Two   Clear the statistical registers.
Step Three Enter the data.
Step Four   Make sure the correct number of scores have been entered.
Step Five   Hit the key that displays the mean.
Step Six   Hit the key that displays the standard deviation.

Note that when using the calculator the standard deviation is found before the variance, which is the opposite from using the definitional formula. The results using the calculator and the definitional formula should agree, within rounding error. In this case rounding error, or imprecision that results from not using the entire number (3.33 rather than 3.3333333...), should be less than one or two hundredths in value.

11-2d Calculating Statistics Using SPSS

More often than not, statistics are computed using a computer package such as SPSS. It may seem initially like a lot more time and trouble to use the computer to do such simple calculations, but you will likely appreciate the savings in time and effort at a later time.

The first step is to enter the data into a form the computer can recognize. A data file with the example numbers is shown in the following figure:

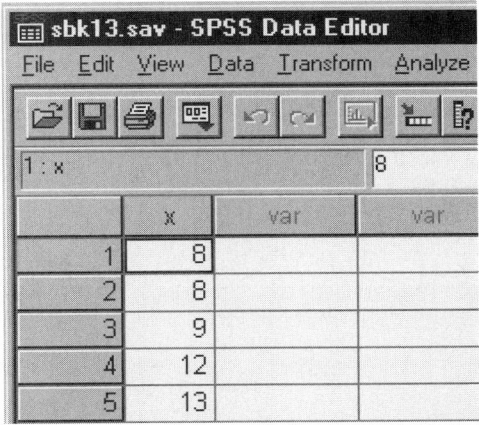

Entering scores in the SPSS data editor.

Any number of statistical commands in SPSS would result in the computation of simple measures of central tendency and variability. The next figure illustrates how to access the Frequencies command—by selecting Analyze/Descriptive Statistics/Frequencies.

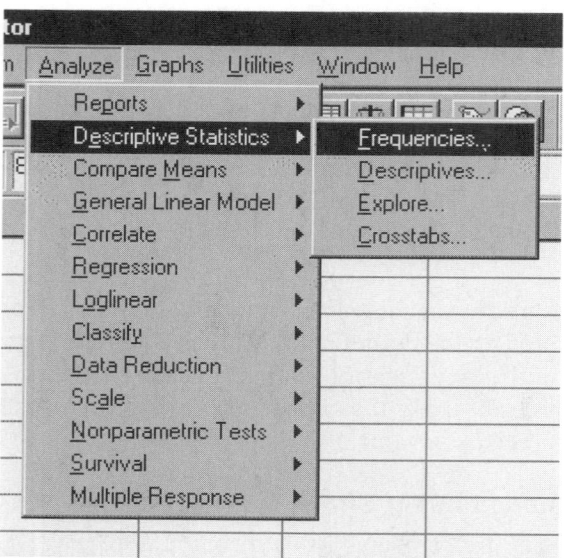

The steps to find statistics using SPSS.

In the Frequencies screen, the variable X is clicked from the left box to the right box, and then the Statistics button on the form is clicked, which opens the Frequencies: Statistics dialog box. The following figure shows which options to select in that dialog box: the four statistics under Central Tendency and the first three under Dispersion.

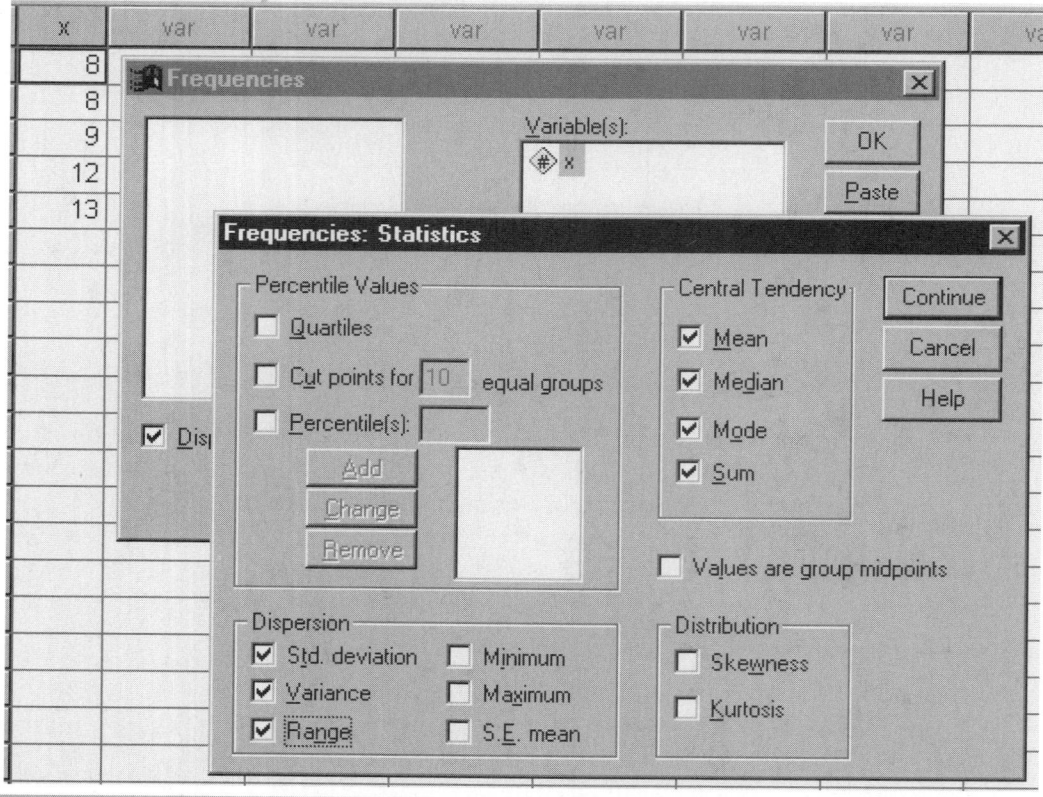

Selecting statistical options when using the Frequencies command in SPSS.

Once those selections are made, the user clicks Continue and then OK, and the results of the procedure appear, as shown in the following figure:

Statistics

X

N	Valid	5
	Missing	0
Mean		10.00
Median		9.00
Mode		8
Std. Deviation		2.35
Variance		5.50
Range		5
Sum		50

Example output from the SPSS Frequencies command.

11-3 Interpreting a Mean and Standard Deviation

An analysis, called a breakdown, gives the means and standard deviations of a variable for each level of another variable. The means and standard deviations may then be compared. If the means are different, then the groups as a whole differ from each other. If the standard deviations differ, it means that the scores in the group with the smaller standard deviation are more similar to each other than the scores in the groups with the larger standard deviation. Going back to the example of shoe sizes, the raw data appeared as follows:

Shoe Size, Shoe Width, and Gender

Shoe Size	Shoe Width	Sex
10.5	B	M
6.0	B	F
9.5	D	M
8.5	A	F
7.0	B	F
10.5	C	M
7.0	C	F
8.5	D	M
6.5	B	F
9.5	C	M
7.0	B	F
7.5	B	F
9.0	D	M
6.5	A	F
7.5	B	F

The corresponding data file in SPSS would appear as follows:

	shoesize	shoewdth	sex
1	10.5	B	0
2	6.0	B	1
3	9.5	D	0
4	8.5	A	1
5	7.0	B	1
6	10.5	C	0
7	7.0	C	1
8	8.5	D	0
9	6.5	B	1
10	9.5	C	0
11	7.0	B	1
12	7.5	B	1
13	9.0	D	0
14	6.5	A	1
15	7.5	B	1

sbk13a.sav - SPSS Data Editor
File Edit View Data Transform Analyze Gra
16 : sex

Entering the data into the data editor of SPSS.

It is possible to compare the shoe sizes of males and females by first finding the mean and standard deviation of males only and then for females only. In order to do this the original shoe sizes would be partitioned into two sets, one of males and one of females, as follows:

Splitting the Scores into Two Groups, Males and Females

Males	Females
10.5	6.0
9.5	8.5
10.5	7.0
8.5	7.0
9.5	6.5
9.0	7.0
	7.5
	6.5
	7.5

The means and standard deviations of the males and females are organized into a table as follows:

Means and Standard Deviations of Shoe Size Broken Down by Gender

Sex	N	Mean	Standard Deviation
Males	6	9.58	0.80
Females	9	7.06	0.73
Total	15	8.06	1.47

It can be seen that the males had larger shoe sizes as evidenced by the larger mean. It can also be seen that males also had somewhat greater variability as evidenced by their larger standard deviation. In addition, the variability within groups (males and females separately) was considerably less than the total variability (both sexes combined).

The analysis just described may be done using the SPSS Means command. The Means command is accessed by selecting Analyze/Compare Means/Means, as shown in the following figure:

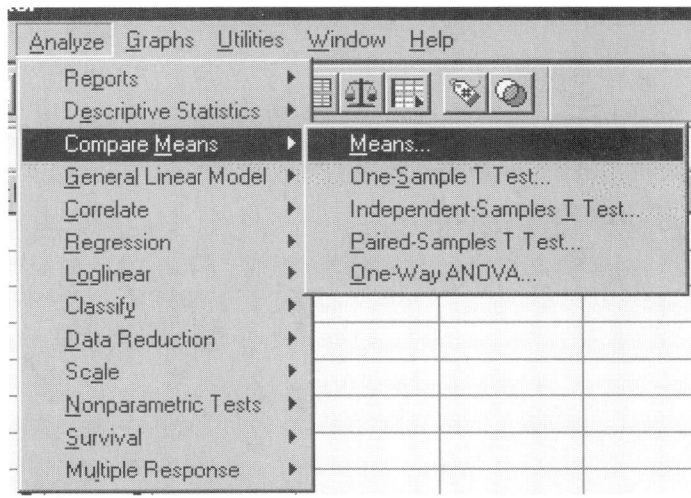

Procedure to find means and standard deviations for separate groups using SPSS.

The Means dialog box appears. It should look like the one in the following figure:

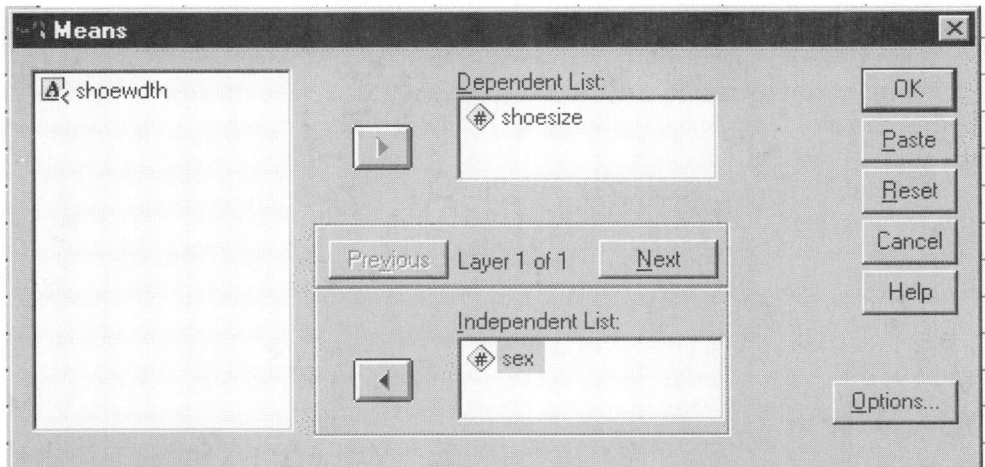

Selecting the dependent and independent variables.

In this example, shoe size is selected as the dependent variable and sex as the independent variable. Means and standard deviations will be found for each variable in the dependent list for each level of each variable in the independent list. The grouping variables will always appear in the independent list, and the measured variables in the dependent list. An easy way to remember this is that the dependent variables depend upon the independent variables. In this case, shoe size depends upon gender.

Clicking the OK button produces the results of this analysis, which are presented in the following figure:

Report

SHOESIZE

SEX	Mean	N	Std. Deviation
Male	9.583	6	.801
Female	7.056	9	.726
Total	8.067	15	1.474

Example output from the SPSS Means command.

A similar kind of breakdown could be performed for shoe size broken down by shoe width, which would produce the following table:

Means and Standard Deviations of Shoe Size Broken Down by Shoe Width			
Shoe Width	N	Mean	Standard Deviation
A	2	7.5	1.41
B	7	7.43	1.46
C	3	9.0	1.80
D	3	9.0	0.50
Total	15	8.06	1.47

A breakdown is a very powerful tool in examining the relationships between variables. It can express a great deal of information in a relatively small space.

Summary

Statistics serve to estimate model parameters and describe the data. Two categories of statistics were described in this chapter: measures of central tendency and measures of variability. In the former category were the mean, median, and mode. In the latter were the range, standard deviation, and variance. Measures of central tendency describe a typical or representative score, while measures of variability describe the spread or dispersion of scores. Both definitional examples of computational procedures and procedures for obtaining the statistics from a calculator were presented.

Chapter

12

Score Transformations

Key Terms

percentile rank based on the
  normal curve

percentile rank based on the
  sample
percentile ranks

raw score
transformation
transformed scores

If a student, upon viewing a recently returned test, found that he or she had made a score of 33, would that be a good score or a poor score? Based only on the information given, it would be impossible to tell. The 33 could be out of 35 possible questions and be the highest score in the class, or it could be out of 100 possible points and be the lowest score, or anywhere in between. The score that is given is called a **raw score.** The purpose of this chapter is to describe procedures to convert raw scores into **transformed scores,** which give meaning to the numbers and allow comparisons between scores made on different scales.

12-1 Why Do We Need to Transform Scores?

Converting scores from raw scores into transformed scores has two purposes:

- It gives meaning to the scores and allows some kind of interpretation of the scores.

- It allows direct comparison of two scores. For example, a score of 33 on the first test might not mean the same thing as a score of 33 on the second test.

The **transformations** discussed in this section belong to two general types; percentile ranks and linear transformations. Percentile ranks are advantageous in that the average person has an easier time understanding and interpreting their meaning. However, percentile ranks also have a rather unfortunate statistical property that makes their use generally unacceptable among the statistically sophisticated. This chapter will focus on percentile ranks. Linear transformations are the topic of Chapter 13.

12-2 Percentile Ranks Based on the Sample

A **percentile rank** is the percentage of scores that fall below a given score. For example, a raw score of 33 on a test might be transformed into a percentile rank of 98 and interpreted as "You did better than 98% of the students who took this test." In that case the student would feel pretty good about the test. If, on the other hand, a percentile rank of 3 was obtained, the student might wonder what he or she was doing wrong.

The procedure for finding the percentile rank is as follows:

1. Rank order the scores from lowest to highest.

2. Find the proportion of scores that fall below the score and convert to a percentage by multiplying by 100.

3. Find one-half the proportion of scores that fall at the score and convert to a percentage by multiplying by 100.

4. Add the percentage of scores that fall below the score to one-half the percentage of scores that fall at the score.

The result is the percentile rank for that score.

It's actually easier to demonstrate and perform the procedure than it sounds. For example, suppose the obtained scores from 11 students were

33 28 29 37 31 33 25 33 29 32 35

1. You want to know the percentile rank for the score of 31. The first step would be to rank order the scores from lowest to highest:

25 28 29 29 31 32 33 33 33 35 37

2. Computing the percentage falling below a score of 31, for example, gives the value 4/11 = .364 or 36.4%. The four in the numerator reflects that four scores (25, 28, 29, and 29) were less than 31. The 11 in the denominator is N, or the number of scores.

3. The percentage falling at a score of 31 would be 1/11 = .0909 or 9.09%. The numerator is the number of scores with a value of 31 (one) and the denominator again being the number of scores (11). One-half of 9.09 would be 4.55.

4. Adding the percentage below the score to one-half the percentage within the score yield a percentile rank of 36.4 + 4.55 or 40.95%.

These computations are illustrated in the following figure:

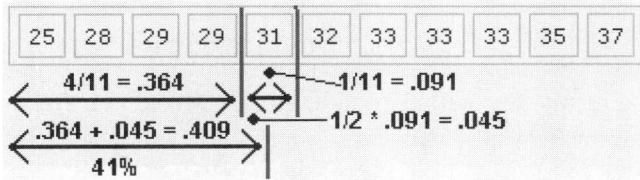

Similarly, for a score of 33, the percentile rank would be computed by adding the percentage below (6/11 = .5454 or 54.54%) to one-half the percentage within (1/2 * 3/11 = .1364 or 13.64%), producing a percentile rank of 68.18%. The 6 in the numerator of percentage below indicates that 6 scores were smaller than a score of 33, while the 3 in the percentage within indicates that 3 scores had the value 33. All three scores of 33 would have the same percentile rank of 68.18%. The following figure shows the computation for the score of 33.

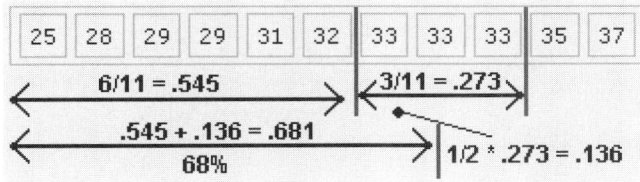

This procedure can be described in an algebraic expression as follows:

$$PR = \left(\frac{f_b + \frac{1}{2} f_w}{N}\right) * 100$$

where:
  f_b is the frequency below; the number of scores which are less than the score value of the percentile rank
  f_w is the frequency within; the number of scores which have the same value as the score value of the percentile rank
  N is the number of scores

Computational formula for percentile ranks based on the sample.

Application of this algebraic procedure to the score values of 31 and 33 would give the following results:

$$PR_{31} = \frac{f_b + \frac{1}{2}f_w}{N} * 100 = \frac{4 + \frac{1}{2} \cdot 1}{11} * 100 = \frac{4.5}{11} * 100 = 40.9\%$$

The percentile rank based on the sample is computed for a score of 31.

$$PR_{33} = \frac{f_b + \frac{1}{2}f_w}{N} * 100 = \frac{6 + \frac{3}{2} \cdot 1}{11} * 100 = \frac{7.5}{11} * 100 = 68.2\%$$

A second example of computing percentile ranks based on the sample, this time for a score of 33.

Note that these results are within rounding error of the percentile rank computed using the procedure described in words.

When computing the percentile rank for the smallest score, the frequency below is zero (0), because no scores are smaller than it. Using the formula to compute the percentile rank of the score of 25:

$$PR_{25} = \frac{f_b + \frac{1}{2}f_w}{N} * 100 = \frac{0 + \frac{1}{2} \cdot 1}{11} * 100 = \frac{0.5}{11} * 100 = 4.6\%$$

An example of computing a percentile rank based on the sample for the lowest score.

Computing the percentile rank for the largest score, 37, gives:

$$PR_{33} = \frac{f_b + \frac{1}{2}f_w}{N} * 100 = \frac{10 + \frac{1}{2} \cdot 1}{11} * 100 = \frac{10.5}{11} * 100 = 95.4\%$$

An example of computing the percentile rank based on the sample for the highest score.

The last two cases demonstrate that a score may never have a percentile rank equal to or less than zero or equal to or greater than 100. Percentile ranks may be closer to zero or one hundred than those obtained if the number of scores increases. The percentile ranks for all the scores in the example data are shown in the following table:

Percentile Ranks Based on the Sample										
Score	25	28	29	29	31	32	33	33	33	35	37
Percentile Rank	4.6	13.6	27.3	27.3	40.9	50	68.2	68.2	68.2	86.4	95.4

12-3 Percentile Ranks Based on the Normal Curve

The percent of area below a score on a normal curve with a given mu and sigma provides an estimate of the percentile rank of a score. The mean and standard deviation of the sample estimate the values of mu and sigma. You can use the Probability Calculator to find percentile ranks with these values.

Here are the sample raw scores for this example:

25   28   29   29   31   32   33   33   33   35   37

The sample mean is 31.364 and the sample standard deviation is 3.414.

To use the Probability Calculator to find percentile rank of a score (29 for this example) based on the normal curve, follow these steps:

1.   Select Area Below under the Normal Curve button, and then click Normal Curve.

2.   Enter the value of the mean (31.364) in the Mu box, the standard deviation (3.414) in the Sigma box, and the score (29) in the Score box.

3.   Click the right-arrow button.

4.   Multiply the Probability that results (.2443) by 100, and put a "%" after it: 24%.

So, entering the appropriate values in the Probability Calculator for a score of 29 yields a percentile rank based on the normal curve of 24%. The following figure illustrates the steps involved.

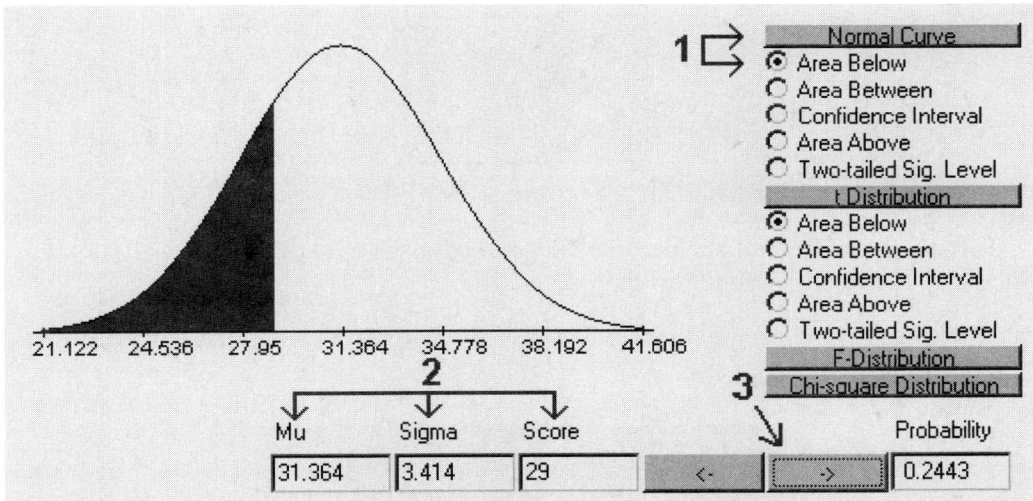

Computing percentile rank based on the normal curve for a score of 29.

Percentile ranks based on normal curve area for all the example scores are shown in the following table:

Percentile Ranks Based on the Normal Curve										
Score	25	28	29	29	31	32	33	33	33	35	37
Percentile Rank	3	16	24	24	46	57	68	68	68	86	95

12-4 Computing Percentile Ranks Based on the Normal Curve with SPSS

Percentile ranks based on normal area can be easily computed using SPSS. The first step is to enter the scores in a variable in a data file. In this example, the variable is labeled x.

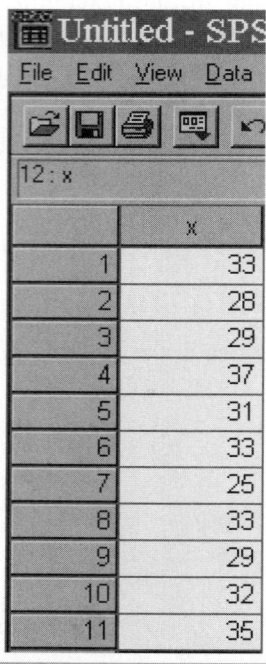

A raw data file in SPSS.

To find the mean and standard deviation (and also to add an additional variable called standard scores to the data file), follow these steps:

1. Choose Analyze/Descriptive Statistics/Descriptives as shown in the following figure:

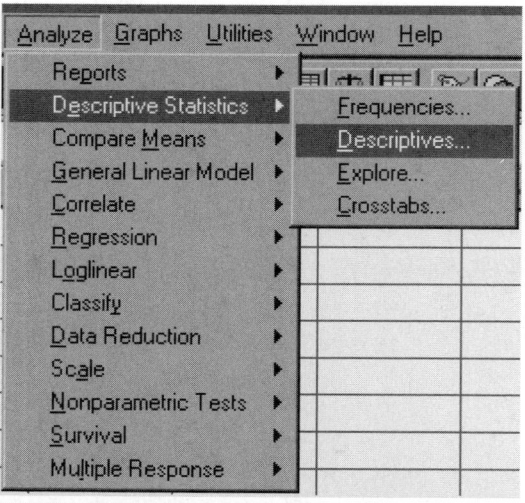

Commands generating descriptive statistics in SPSS.

2. The Descriptives dialog box appears. The variable x will appear in the left box; click the directional button between the boxes to move x to the right box. Your screen should now look like the following figure. Note that the Save standardized values as variables box has been checked.

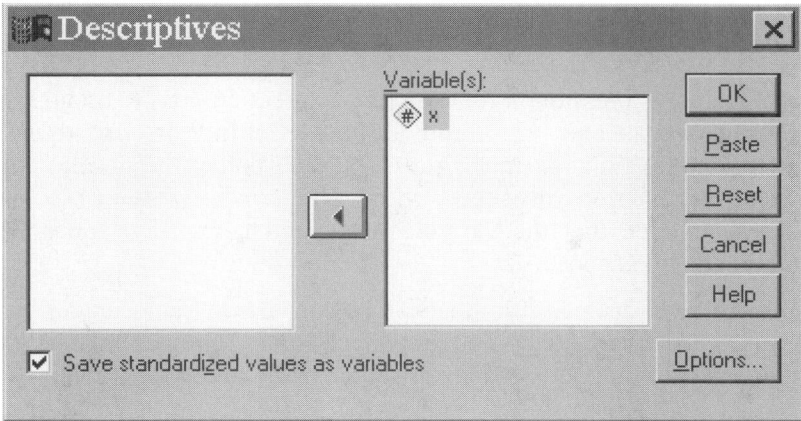

The Descriptives command in SPSS.

3. When you are ready, click the OK button. This command produces two results:

  a. A table of means and standard deviations for the x variable in the output screen.

  b. The addition of a second variable, called zx, in the data table. In general, the new variable name will be the original variable with a "z" in front of it.

The data file should now look like the one in the following figure:

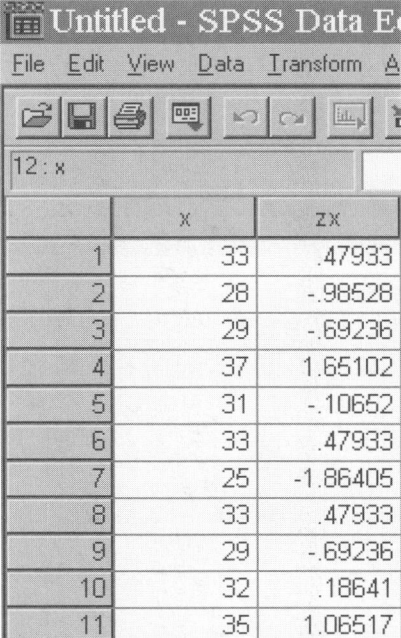

The SPSS data file showing standard scores after the Descriptives command has been executed.

4. The next step is to find the area that falls below the value of x on a normal curve with mu and sigma equal to the mean and standard deviation of the scores, respectively. This is done in SPSS by means of the Compute command (select Transform/Compute).

5. Enter a name for the variable to be computed in the Target Variable box. In this example, the name is prnormal (a shortened form of percentile rank using the normal distribution).

6. Place an algebraic expression in the Numeric Expression box by finding the function you want in the Functions list, and double-clicking it. In this example, the CDFNORM(zvalue) function that returns the area below the normal curve is selected. SPSS places the function you choose in the Numeric Expression box, as the following figure shows. Notice that the variable just created, zx, now appears in parentheses following function you chose.

7. The result will be in decimal form that can be converted to percentages by multiplying by 100. Be sure to move the * 100 outside of the right parenthesis in the Numeric Expression box: It should look exactly as shown in the following figure:

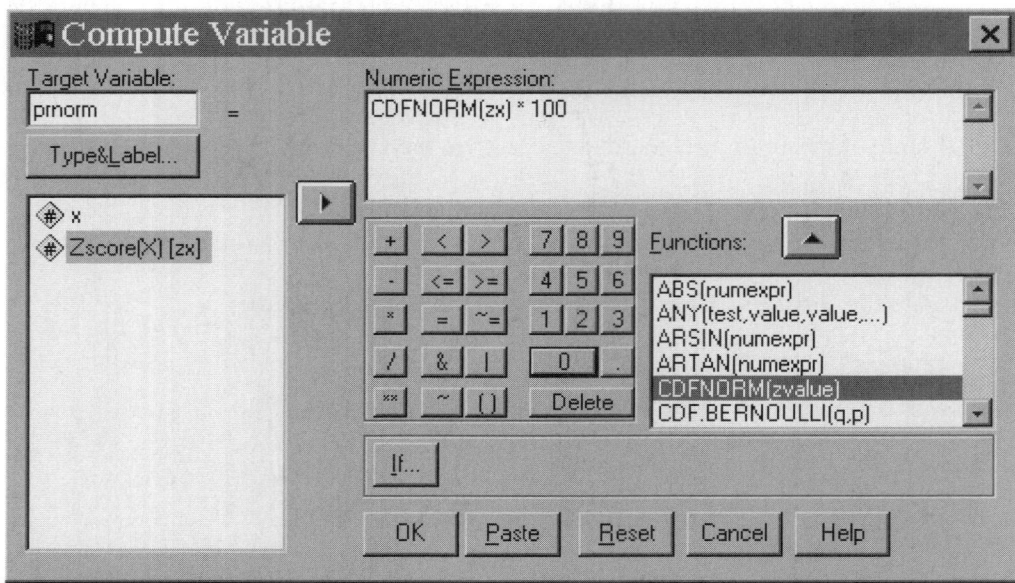

Computing percentile ranks based on the normal curve using SPSS.

8. Click OK and all values of zx are converted to a value of prnormal.

The result is a new variable, called prnormal (prnorm), that is included in the data table that follows.

Untitled - SPSS Data Editor

File Edit View Data Transform Analyze Graph:

12 : x

	x	zx	prnorm
1	33	.47933	68.41
2	28	-.98528	16.22
3	29	-.69236	24.44
4	37	1.65102	95.06
5	31	-.10652	45.76
6	33	.47933	68.41
7	25	-1.86405	3.12
8	33	.47933	68.41
9	29	-.69236	24.44
10	32	.18641	57.39
11	35	1.06517	85.66

Results of computing percentile ranks based on the normal curve using SPSS.

12-5 Comparing the Two Methods of Computing Percentile Ranks

The astute student will observe that the percentile ranks based on the normal curve are somewhat different from those called percentile ranks based on the sample. That is because the two procedures give percentile ranks that are interpreted somewhat differently. The following table shows some example raw scores and their percentile ranks computed two different ways.

Comparing the Two Methods of Computing Percentile Ranks										
Raw Score	25	28	29	29	31	32	33	33	33	35	37
Sample %ile	4.6	13.6	27.3	27.3	40.9	50	68.2	68.2	68.2	86.4	95.4
Normal Area %ile	3	16	24	24	46	57	68	68	68	86	95

The **percentile rank based on the sample** describes where a score falls relative to the scores in the sample distribution. That is, if a score has a percentile rank of 34 using this procedure, then it can be said that 34% of the scores in the sample distribution fall below it.

The **percentile rank based on the normal curve,** on the other hand, describes where the score falls relative to a hypothetical model of a distribution. That is a score with a percentile rank of 34 using the normal curve says that 34% of an infinite number of scores obtained using a similar method will fall below that score. The additional power of this last statement is not bought without cost, however, in that the assumption must be made that the normal curve is an accurate model of the sample distribution, and that the sample mean and standard deviation are accurate estimates of the model parameters mu and sigma. If you accept these assumptions, then the percentile rank based on normal area describes the relative standing of a score within an infinite population of scores.

12-6 An Unfortunate Property

Percentile ranks, as the name implies, is a system of ranking. Using the system destroys the interval property of the measurement system. That is, if the scores could be assumed to have the interval property before they were transformed, they would not have the property after transformation. The interval property is critical to interpreting most of the statistics described in this text—such as mean, standard deviation, and variance—thus transformation to percentile ranks does not permit meaningful analysis of the transformed scores.

If an additional assumption of an underlying normal distribution is made, not only do percentile ranks destroy the interval property, but they also destroy the information in a particular manner. If the scores are distributed normally, then percentile ranks underestimate large differences in the tails of the distribution and overestimate small differences in the middle of the distribution. This is most easily understood in an illustration:

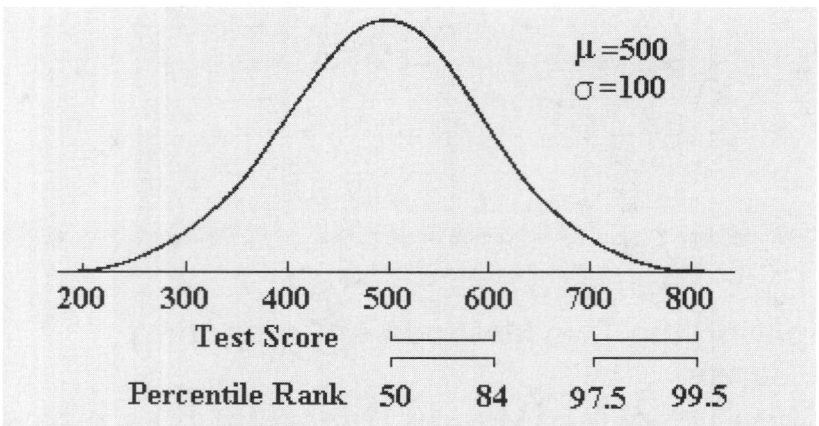

Distortion of percentile ranks when the underlying distribution is a normal curve.

In the example depicted in the illustration, two standardized achievement tests with $\mu = 500$ and $\sigma = 100$ were given. In the first, an English test, Suzy made a score of 500 and Johnny made a score of 600, thus there was a 100-point difference between their raw scores. On the second, a math test, Suzy made a score of 800 and Johnny made a score of 700, again a 100-point difference in raw scores. It can be said, then, that the differences on the scores on the two tests were equal, one hundred points each.

When converted to percentile ranks, however, the differences are no longer equal. On the English test, Suzy receives a percentile rank of 50 while Johnny gets an 84, a difference of 34 percentile rank points. On the math test, Johnny's score is transformed to a percentile rank of 97.5 while Suzy's percentile rank is 99.5, a difference of only two percentile rank points.

So you can see that a percentile rank has a different meaning depending upon whether it occurs in the middle of the distribution or the tails of a normal distribution. Differences in the middle of the distribution are magnified, and differences in the tails are minimized.

The unfortunate property destroying the interval property precludes the use of percentile ranks by sophisticated statisticians. Percentile ranks will remain in widespread use in order to interpret scores to the layman, but the statistician must help in emphasizing and interpreting scores. Because of this unfortunate property, a different type of transformation is needed, one that does not destroy the interval property. This leads directly into the topic of the next chapter, "Linear Transformations."

Summary

Transformed scores make meaningful sense out of raw scores, allowing for interpretation and comparison of the numbers. There are two general types of transformations: percentile ranks and linear transformations. This chapter discussed percentile ranks in detail.

A percentile rank based on the sample is the percentage of scores that fall below a given score. Converting a raw test score to a percentile rank based on the sample, one could say that Maryann, for example, scored better than 78% of the students who took the test.

A percentile rank based on the normal curve is the percentage of scores that fall below a hypothetical distribution of scores. Converting a raw test score to a percentile rank based on the normal curve, one could say that Maryann scored better than 83% of the students who would ever take the test. This powerful statement can only be made if certain assumptions are true, namely that the true distribution is a normal curve and that the sample mean and standard deviation are accurate estimates of the true parameters mu and sigma of the normal curve.

Both the percentile rank based on the sample and the percentile rank based on the normal curve destroy the interval property of the measurement system. If the interval property was satisfied before the transformation, it will not be so after the transformation. If the underlying distribution is a normal curve, the interval property is destroyed in a particular manner, with percentile ranks near the middle of the distribution meaning less change than percentile ranks in the tails of the distribution.

Chapter

13

Linear Transformations

Key Terms

additive component
intercept
IQ scale

linear transformations
multiplicative component
slope

standard scores
stanine transformation
T score

A **linear transformation** is a transformation of the form $X' = a + bX$. If a measurement system approximated an interval scale before the linear transformation, it will approximate it to the same degree after the linear transformation. Other properties of the distribution are similarly unaffected. For example, if a distribution was positively skewed before the transformation, it will be positively skewed after.

The symbols in the transformation equation ($X' = a + bX_i$) have the following meanings:

Symbol	Meaning
X'_i	Score after the transformation (read X prime or X transformed)
a	Additive component (sometimes called the **intercept**)
b	Multiplicative component (sometimes called the **slope**)
X_i	Raw score

The a and b of the transformation are set to real values to specify a transformation.

You perform the transformation by first multiplying each score value by the multiplicative component (b), and then adding the additive component (a) to it. The following set of data is linearly transformed with the transformation $X'_i = 20 + 3 * X_i$ (that is, $a = 20$ and $b = 3$).

Linear Transformation Where a = 20, b = 3	
X	X' = a + bX
12	56
15	65
15	65
20	80
22	86

The score value of 12, for example, was transformed first by multiplying by 3 ($12 * 3 = 36$), and then by adding 20 to that result to produce 56.

The effect of the linear transformation on the mean and standard deviation of the scores is of considerable interest. For that reason, both the additive and multiplicative components of the transformation will be examined separately for their relative effects.

13-1 The Additive Component

If the multiplicative component is set equal to one, the linear transformation becomes $X' = a + X$, so that the effect of the **additive component** may be examined. In this transformation, a constant is added to every score. An example additive transformation is shown in the following table:

Linear Transformation Where a = 20, b = 1			
	X		X' = a + bX
	12		32
	15		35
	15		35
	20		40
	22		42
$\bar{X} =$	16.8	$\bar{X}' =$	36.8
$s_X =$	4.09	$s_{X'} =$	4.09

The transformed mean, $\bar{X}'$, is equal to the original mean, $\bar{X}$, plus the transformation constant, in this case 20 (a = 20). The standard deviation (s_x) does not change. The mean and standard deviation are "transformed" as shown in the following equations:

$$\bar{X}' = a + \bar{X}$$

$$s_{X'} = s_X$$

That is, the transformed mean is equal to the original mean plus the additive component, and the transformed standard deviation is equal to the original standard deviation. It is as if the distribution was lifted up and placed back down to the right or left, depending upon whether the additive component was positive or negative. Here's a graphical look at the effect of the additive component:

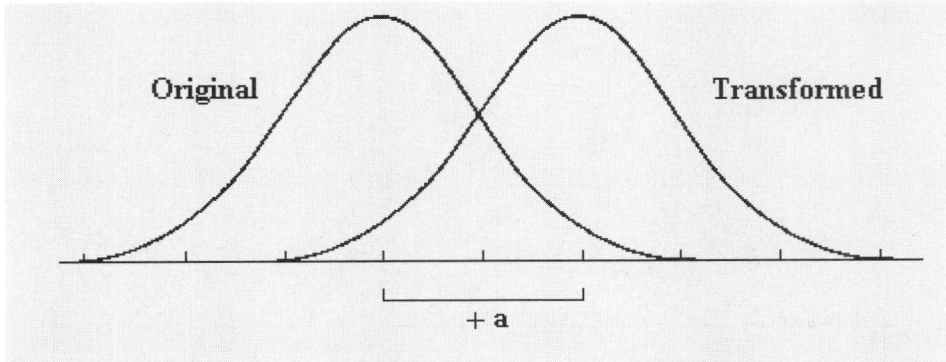

The additive transformation changes the midpoint of the distribution without changing the shape of the distribution.

13-2 The Multiplicative Component

The effect of the **multiplicative component** (b) may be examined separately if the additive component is set equal to zero. The transformation equation becomes X' = bX, which is the type of transformation done when the scale is changed, for example from feet to inches. In that case, the value of b would be equal to 12 because there are 12 inches to the foot. Similarly, transformations to and from the metric system, such pounds to kilograms and back again, are multiplicative transformations.

An example multiplicative transformation, with b = 3, is presented in the following table:

Linear Transformation Where a = 0, b = 3		
	X	X' = a + bX
	12	36
	15	45
	15	45
	20	60
	22	66
$\bar{X}$ =	16.8	$\bar{X}'$ = 50.4
s_X =	4.09	$s_{X'}$ = 12.26

Note that both the mean and the standard deviation of the transformed scores are three times their original value, which is precisely the amount of the multiplicative component. The mean and standard deviation are changed as shown in the equations below:

$$\overline{X}' = b\,\overline{X}$$
$$s_{X'} = bs_X$$

That is, the transformed mean is equal to the multiplicative component times the original mean, and the transformed standard deviation is equal to the multiplicative component times the original standard deviation. The multiplicative component, then, affects both the mean and standard deviation by its size, as illustrated graphically here:

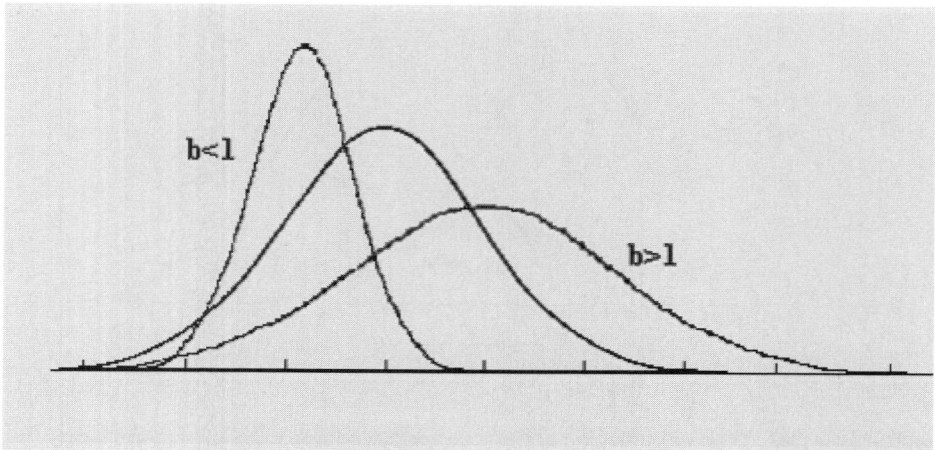

Effect of different multiplicative constants on the shape and midpoint of the distribution.

13-3 Linear Transformations: Effect on Mean and Standard Deviation

Putting the separate effects of the additive and multiplicative components together in a linear transformation, it would be expected that the standard deviation would be affected only by the multiplicative component, and the mean by both. Again, the following two equations express the relationships:

$$\overline{X}' = a + b\,\overline{X}$$
$$s_{X'} = bs_X$$

Effect of a linear transformation on the mean and standard deviation.

For example, in the original distribution (mean 16.8 and standard deviation 4.09) and the linear transformation of X' = a + b * X = 20 + 3 * X, the transformed mean and standard deviation would be expected to be:

$$\bar{X}' = a + b\,\bar{X}$$

$$\bar{X}' = 20 + 3 * 16.8 = 70.4$$

$$s_{X'} = b s_X$$

$$s_{X'} = 3 * 4.09 = 12.26$$

Example of the effect of a linear transformation on the mean.

If the transformation is done and the new mean and standard deviation computed, this is exactly what is found, within rounding error. The following table shows the original and transformed scores, means, and standard deviations.

Linear Transformation Where a = 20, b = 3		
X		X' = a + bX
12		56
15		65
15		65
20		80
22		86
$\bar{X}$ = 16.8	$\bar{X}'$ =	70.4
s_X = 4.09	$s_{X'}$ =	12.26

The question students most often ask at this point is, "Where do the values of a and b come from?" "I am just making them up as I go along." is my most common response. That is exactly what has been done up to this point. Now let's look at the procedure for finding a and b such that the new mean and standard deviation will be a given value.

13-4 Linear Transformations: Finding a and b, Given $\bar{X}'$ and s_X

Suppose that the original scores were raw scores from an intelligence test. Historically, IQs or intelligence test scores have a mean of 100 and a standard deviation of either 15 or 16, depending upon the test selected. In order to convert the raw scores to IQ scores on an **IQ scale**, a linear transformation is performed such that the transformed mean and standard deviation are 100 and 16, respectively. The problem is summarized in the following table:

Linear Transformation Where a and b Are Unknown		
X		X' = a + bX
12		?
15		?
15		?
20		?
22		?
$\bar{X}$ = 16.8	$\bar{X}'$ =	100.0
s_X = 4.09	$s_{X'}$ =	16.0

This problem may be solved by first recalling how the mean and standard deviation are affected by the linear transformation:

$$\overline{X}' = a + b\,\overline{X}$$

$$s_{X'} = bs_X$$

Effect of a linear transformation of the mean and standard deviation.

In these two equations are two unknowns, a and b. Because these equations are independent, they may be solved for a and b in the following manner. First, solving for b by dividing both sides of the equation by s_X produces:

$$s_{X'} = bs_X$$

$$b = \frac{s_{X'}}{s_X}$$

Solving for the value of b.

Thus the value of b is found by dividing the new or transformed standard deviation by the original standard deviation.

After finding the value of b, the value of a may be found in the following manner:

$$\overline{X}' = a + b\,\overline{X}$$

$$\overline{X}' - b\,\overline{X} = a + b\,\overline{X} - b\,\overline{X}$$

$$\overline{X}' - b\,\overline{X} = a$$

$$a = \overline{X}' - b\,\overline{X}$$

Solving for the value of a.

The product of the value of b times the original mean is subtracted from the new or transformed mean.

These two equations can be summarized as follows:

$$b = \frac{s_{X'}}{s_X}$$

$$a = \overline{X}' - b\,\overline{X}$$

Solving for both the multiplicative and the additive constants in a linear transformation.

Application of these equations to the original problem where $\overline{X}' = 100$ and $s_{X'} = 16$ produces the following results:

$$b = \frac{s_{X'}}{s_X} = \frac{16}{4.09} = 3.91$$

$$a = \overline{X}' - b\,\overline{X} = 100 - (3.91 * 16.8) = 34.31$$

An example of applying the formulas to solve for linear transformation constants.

Plugging these values into the original problem produces the desired results, within rounding error, as the following table shows:

Linear Transformation Where a = 34.31, b = 3.91			
	X		X' = a + bX
	12		81.23
	15		92.96
	15		92.96
	20		112.51
	22		120.36
$\bar{X} =$	16.8	$\bar{X}' =$	100.04
$s_X =$	4.09	$s_{X'} =$	15.99

The transformed scores are now on an IQ scale. Before they may be considered as IQs, however, the test must be validated as an IQ test. If the original test had little or nothing to do with intelligence, then the IQs that result from a linear transformation such as this one would be meaningless. Just because a statistician can transform any test into an IQ scale does not make that test an IQ test.

Note: The values for the transformed mean (100) and standard deviation (16) for an IQ scale can be traced historically. Originally, intelligence was calculated by dividing the mental age of the subject by the chronological age, and then multiplying by 100. The mental age was the average age where the norm group of a given chronological age was unable to perform the task. Calculating IQs using this procedure resulted in an average IQ of 100, thus the mean of 100 for the IQ scale. The standard deviations were different depending upon the age of the subject, so that the meaning of an IQ of 113 might depend upon the age of the subject. Transformation IQs standardized the standard deviation at 15 or 16, depending upon the test. With a transformation IQ, an IQ of 113 means the same thing no matter what the age of the subject. When the IQ test constructors changed from ratio IQs to transformation IQs they could have selected any new mean and standard deviation they wanted, but they chose to keep the new system as similar to the old as possible because psychologists already understood how to interpret the old system.

Using the previous procedure, a given distribution with a given mean and standard deviation may be transformed into another distribution with any given mean and standard deviation. In order to turn this flexibility into some kind of order, some kind of standard scale has to be selected. The IQ scale is one such standard, but its use is pretty well limited to intelligence tests. Another standard is the **T score** (called "capital T" scores), where scores are transformed into a scale with a mean of 50 and a standard deviation of 10. This transformation has the advantage of always being positive and between the values of 1 and 100. Another transformation is the **stanine transformation** where scores are transformed to a distribution with a mean of 5 and a standard deviation of 2. In this transformation the decimals are dropped, so a score of an integer value between 1 and 9 is produced. The Army used this transformation because the results could fit in a single column on a computer card.

13-5 Standard Score or Z-Scores

Another possible transformation is so important and widely used that it deserves an entire section to itself. It is the **standard score** or z-score transformation. The standard score transformation is a linear transformation such that the transformed mean and standard deviation are 0 and 1 respectively. The selection of these values was somewhat arbitrary, but not without some reason.

Transformation to z-scores could be accomplished using the procedure described earlier to convert any distribution to a distribution with a given mean and standard deviation, in this case 0 and 1. Here's how to solve for a and b from the example data:

$$b = \frac{s_{X'}}{s_X} = \frac{1}{4.09} = 0.244$$

$$a = \overline{X}' - b\,\overline{X} = 0 - (0.244 * 16.8) = -4.11$$

Example of solving for the transformational constants to do a standard score transformation.

So for this transformation, a = -4.11 and b = 0.244. The following table shows the transformed scores.

Linear Transformation Where a = -4.11, b = 0.244			
	X		X' = a + bX
	12		-1.18
	15		-0.45
	15		-0.45
	20		0.77
	22		1.26
$\overline{X} =$	16.8	$\overline{X}' =$	-0.01
$s_X =$	4.09	$s_{X'} =$	0.997

Note that the transformed mean and standard deviation (-0.01 and 0.997) are within rounding error of the desired figures (0 and 1).

Using a little algebra, computational formulas to convert raw scores to z-scores may be derived. When converting to standard scores ($\overline{X}' = 0$ and $s_{X'} = 1.0$), the values of b and a can be found starting with the general formulas for finding b and a and substituting the known values of the new mean and standard deviation. The values for b and a thus become:

$$b = \frac{s_{X'}}{s_X} = \frac{1}{s_X}$$

$$a = \overline{X}' - b\,\overline{X} = 0 - \frac{1}{s_X} * \overline{X} = \frac{-\overline{X}}{s_X}$$

Solving for the transformational constants of a standard score transformation.

The value for X' can then be found by substituting these values into the linear transformation equation:

$$X' = a + bX$$

$$X' = \left(-\frac{\bar{X}}{s_X}\right) + \left(\frac{1}{s_X} * X\right)$$

$$X' = \frac{X}{s_X} - \frac{\bar{X}}{s_X}$$

$$X' = \frac{X - \bar{X}}{s_X}$$

Deriving a general formula to find standard scores.

The last result is a computationally simpler version of the standard score transformation. All this algebra was done to demonstrate that the standard score or z-score transformation was indeed a type of linear transformation. (If you are unable to follow the mathematics underlying the algebraic transformation, you will just have to "Believe!") In any case, the formula for converting to z-scores is

$$z = \frac{X - \bar{X}}{s_X}$$

The z-score formula.

(Note that the "z" has replaced the X'.)

Application of this computational formula to the example data yields:

Transformation Using the Z-Score Formula			
	X		z
	12		-1.17
	15		-0.44
	15		-0.44
	20		0.78
	22		1.27
$\bar{X} =$	16.8	$\bar{X}' =$	0.0
$s_X =$	4.09	$s_{X'} =$	.997

The procedure for solving for b and a and the z-score computation formula produce almost identical results, except that the computational formula is slightly more accurate. Because of the increased accuracy and ease of computation, it is the method of choice for finding z-scores.

13-6 Using SPSS for Linear Transformations

All of the transformations described in this chapter could be done using the SPSS Compute command.

Start with the data file in the following figure:

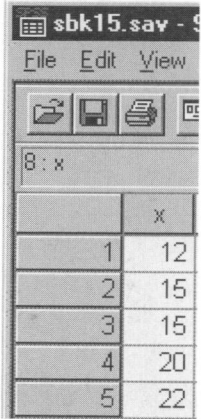

SPSS data file for linear transformations.

The following are a series of Compute commands to perform all the transformations described in this chapter.

To calculate a linear transformation, the Target Variable is lintrans, and you insert the formula into the Numeric Expression box, as this figure shows.

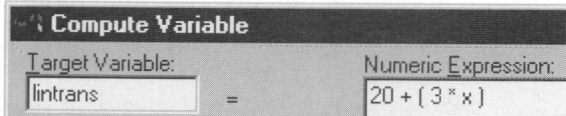

Using the SPSS Compute command to calculate a linear transformation.

In a similar manner, the additive, multiplicative, IQ, and z-score linear transformations could be done. This table lists the Target Variables and the Numeric Expression formulas for a number of them.

Compute Commands for Linear Transformations	
Target Variable	**Numeric Expression**
multrans	$3 * x$
addtrans	$20 + x$
lintrans	$20 + (3 * x)$
IQtrans	$34.31 + (3.91 * x)$
Z1trans	$-4.11 + (0.244 * x)$
Z2Trans	$(x - 16.8)/4.09$
PRnormal	CDFNORM(Z2Trans) * 100

SPSS allows a shortcut to compute the z-scores using the Descriptives command. Select Analyze/Descriptive Statistics/Descriptives as the following figure shows.

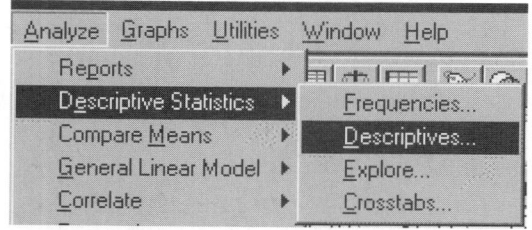

The DESCRIPTIVES command in SPSS

In the command interface that follows, select the variable x and check the Save standardized values as variables box (see the following figure):

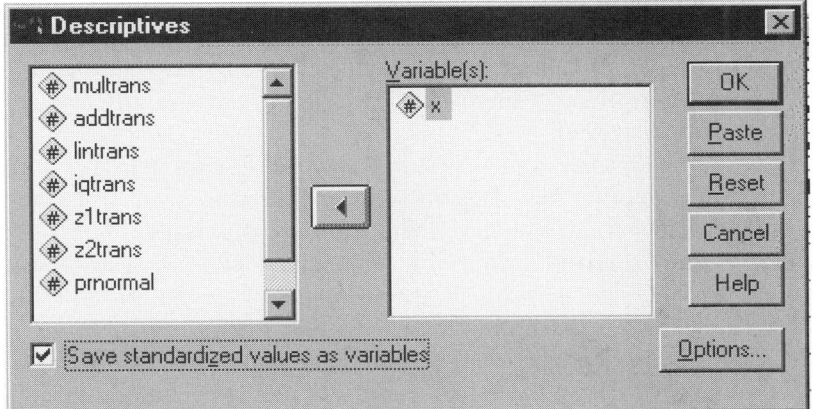

Computing z-scores using the SPSS Descriptives command.

Click OK and this command adds a variable to the data editor that is named with the variable name preceded with a "z." In this case, the new variable would be named zx.

Here's how the variables that result appear in the SPSS Data Editor:

x	multrans	addtrans	lintrans	iqtrans	z1trans	z2trans	prnormal	zx
12	36	32	56	81.23	-1.18	-1.17	12.03	-1.17458
15	45	35	65	92.96	-.45	-.44	32.99	-.44047
15	45	35	65	92.96	-.45	-.44	32.99	-.44047
20	60	40	80	112.51	.77	.78	78.30	.78305
22	66	42	86	120.33	1.26	1.27	89.82	1.27246

Example SPSS data showing the result of various linear transformation.

The means and standard deviations of this file were found using the Descriptives command. Here's the output:

Descriptive Statistics

	N	Mean	Std. Deviation
X	5	16.80	4.09
MULTRANS	5	50.40	12.26
ADDTRANS	5	36.80	4.09
LINTRANS	5	70.40	12.26
IQTRANS	5	99.9980	15.9785
Z1TRANS	5	-.0108	.9971
Z2TRANS	5	.0000	.9992
PRNORMAL	5	49.2272	33.1812
Zscore(X)	5	.0000000	1.0000000
Valid N (listwise)	5		

SPSS output showing the means and standard deviations of the transformed variables.

Summary

Transformations are performed to interpret and compare raw scores. Of the two types of transformations described in this text, percentile ranks are preferred to interpret scores to the lay public, because they are more easily understood. Because of the unfortunate property of destroying the interval property of the scale, the statistician uses percentile rank transformations with reluctance, preferring linear transformations, in which the interval property of the measurement system is not disturbed.

Using a linear transformation, a distribution with a given mean and standard deviation may be transformed into another distribution with a different mean and standard deviation. Several standards for the mean and standard deviation were discussed, but standard scores or z-scores are generally the preferred transformation. The z-score transformation is a linear transformation with a transformed mean of 0 and standard deviation of 1.0. Computational procedures were provided for this transformation.

Standard scores could be converted to percentile ranks by use of the standard normal curve tables. Computation of percentile ranks using this method require additional assumptions about the nature of the world and have the same unfortunate property as percentile ranks based on the sample.

Chapter

14

Regression Models

Key Terms

absolute value
bivariate data
computational formula for the
    standard error of estimate
conditional distribution
confidence interval
form-board test
interval estimate
inverse relationship

least-squares criterion
linear regression
nonoptimal regression model
optimal regression model
point estimate
predicted variable
predictor variable
regression analysis
regression coefficients

regression line
regression model
residuals
scatterplot
standard error of estimate
sum of squared deviations
vectors

Regression models are used to predict one variable from one or more other variables. Regression models provide the scientist with a powerful tool, allowing predictions about past, present, or future events to be made with information about past or present events. The scientist employs these models either because it is less expensive in terms of time and/or money to collect the information to make the predictions than to collect the information about the event itself, or, more likely, because the event to be predicted will occur in some future time. Before describing the details of the modeling process, however, some examples of the use of regression models will be presented.

14-1 Example Uses of Regression Models

Regression models are used for any number of purposes. They can help schools predict which students may need extra help—or extra challenges—in the classroom; assist medical professionals in anticipating problems their patients may develop; aid businesses in predicting which applicants would best help their companies; and help scientists predict earthquakes, to name just a few. Here are some examples of regression models at work.

14-1a Selecting Colleges

A high school student discusses plans to attend college with a guidance counselor. The student has a 2.04 grade-point average out of a maximum 4.00, and mediocre to poor scores on the ACT. He asks about attending Harvard. The counselor tells him he would probably not do well at that institution, predicting he would have a grade point average of 0.64 at the end of four years at Harvard. The student inquires about the necessary grade-point average to graduate and when told that it is 2.25, the student decides that maybe another institution might be more appropriate in case he becomes involved in some "heavy-duty partying."

When asked about the large state university, the counselor predicts that he might succeed, but chances for success are not great, with a predicted grade-point average of 1.23. A regional institution is then proposed, with a predicted grade-point average of 1.54. Deciding that is still not high enough to graduate, the student decides to attend a local community college, graduates with an associate's degree and makes a fortune selling real estate.

If the counselor was using a regression model to make the predictions, he or she would know that although the regression model made very specific predictions, this particular student would probably *not* make a grade point of 0.64 at Harvard, 1.23 at the state university, and 1.54 at the regional university. These values are just "best guesses." It may be that this particular student was completely bored in high school, didn't take the standardized tests seriously, would become challenged in college, and would succeed at Harvard. The selection committee at Harvard, however, when faced with a choice between a student with a predicted grade-point of 3.24 and one with 0.64 would most likely make the rational decision and choose the most promising student.

14-1b Pregnancy

A woman in the first trimester of pregnancy has a great deal of concern about the environmental factors surrounding her pregnancy and asks her doctor about what impact they might have on her unborn child. The doctor makes a "point estimate" based on a regression model that the child will have an IQ of 75. It is highly unlikely that her child will have an IQ of exactly 75, as there is always error in the regression procedure. Error may be incorporated into the information given the woman in the form of an "interval estimate." For example, the doctor could have said that the child had a ninety-five percent chance of having an IQ between 70 and 80; a ninety-five percent chance of an IQ between 50 and 100 would probably also be true, but the range is so great as to not be beneficial to the woman's need for information. The concept of error in prediction is an important part of the discussion of regression models.

It is also worth pointing out that regression models do not make decisions for people. Regression models are a source of information about the world. In order to use them wisely, it is important to understand how they work.

14-1c Selection and Placement during the World Wars

Technology helped the United States and her allies to win the first and second world wars. One usually thinks of the atomic bomb, radar, bombsights, better-designed aircraft, and so forth when this statement is made. Less well known were the contributions of psychologists and associated scientists to the development of test and linear regression prediction models used for selection and placement of men and women in the armed forces.

During these wars, the United States had thousands of men and women enlisting or being drafted into the military. These individuals differed in their ability to perform physical and intellectual tasks. The problem was one of both selection—who is enlisted and who is rejected—and placement of those selected—who will cook and who will fight. The army that takes its best and brightest men and women and places them in the front lines digging trenches is less likely to win the war than the army who places these men and women in positions of leadership.

It costs a great deal of money and time to train a person to fly an airplane. Every time one crashes, the military has lost a plane as well as the time and effort to train the pilot, not to mention the loss of the life of a person. For this reason it was, and still is, vital that the best possible selection and prediction tools be used for personnel decisions.

14-1d Manufacturing Widgets

A new plant to manufacture widgets is being located in a nearby community. The plant personnel officer advertises the employment opportunity and the next morning has 10,000 people waiting to apply for the 1,000 available jobs. It is important to select the 1,000 people who will make the best employees because training takes time and money and firing is difficult and bad for community relations. In order to provide information to help make the correct decisions, the personnel officer employs a regression model.

In order to construct a regression model it is necessary to have information on both the variable we are predicting (the dependent variable) and the variable we are predicting from (the independent variable). This information is used to construct a prediction model that is used in the future. An understanding of the procedure used to create regression models is essential in order to answer the question, "Why do we need to predict the number of widgets made if we already know how many widgets were made?"

14-2 Procedure for Construction of a Regression Model

In order to construct a regression model, both the information that is going to be used to make the prediction and the information that is to be predicted must be obtained from a sample of objects or individuals. The relationship between the two pieces of information is subsequently modeled with a linear transformation. Then in the future, only the first information is necessary, and the regression model is used to transform this information into the prediction. In other words, it is necessary to have information on both variables before the model can be constructed.

For example, the personnel officer of the widget manufacturing company might give all applicants a test and predict the number of widgets made per hour on the basis of the test score. In order to create a regression model, the personnel officer would first have to give the test to a sample of applicants and hire all of them. Later, when the number of widgets made per hour had stabilized, the personnel officer could create a prediction model to predict the widget production of future applicants. All future applicants would be given the test, and hiring decisions would be based on test performance.

A notational scheme is now necessary to describe the procedure:

X_i is the variable used to predict, and is sometimes called the independent variable. In the case of the widget manufacturing example, it would be the test score.

Y_i is the observed value of the **predicted variable,** and is sometimes called the dependent variable. In the example, it is the number of widgets produced per hour by that individual.

Y'_i is the predicted value of the dependent variable. In the example, it is the predicted number of widgets per hour by that individual.

The goal in the regression procedure is to create a model where the predicted and observed values of the variable to be predicted are as similar as possible. For example, in the widget manufacturing situation, we want the predicted number of widgets made per hour to be as close to observed values as possible. The more similar these two values, the better the model. The next section presents a method of measuring the similarity of the predicted and observed values of the predicted variable.

14-3 The Least-Squares Criteria for Goodness-of-Fit

In order to develop a measure of how well a model predicts the data, it is valuable to present an analogy of how to evaluate predictions. Suppose there were two interviewers, Mr. A and Ms. B, who separately interviewed each applicant for the widget manufacturing job for ten minutes. At the end of that time, the interviewer had to make a prediction about how many widgets that applicant would produce two months later. All of the applicants interviewed were hired, regardless of the predictions, and at the end of the two-month trial period, the best interviewer was to be retained and promoted, and the other was to be fired. The purpose of the following is to develop a measure of goodness-of-fit, or how well the interviewer predicted.

The notational scheme for the table is as follows:

Y_i is the observed or actual number of widgets made per hour

Y'_i is the predicted number of widgets

Suppose the data for the five applicants were as shown in this table:

Predictions of Two Interviewers		
	Interviewer	
Observed	Mr. A	Ms. B
Y_i	Y'_i	Y'_i
23	38	21
18	34	15
35	16	32
10	10	8
27	14	23

Obviously neither interviewer was impressed with the fourth applicant, for good reason. A casual comparison of the two columns of predictions with the observed values leads one to believe that interviewer B made the better predictions. A procedure is needed to provide a measure, or single number, of how well each interviewer performed.

The first step is to find by how much each interviewer missed the predicted value for each applicant. This is done by finding the difference between the predicted and observed values for each applicant for each interviewer. These differences are called **residuals.** If the column of differences between the observed and predicted is summed,

$$\sum_{i=1}^{N} (Y_i - Y_i')$$

then it would appear that interviewer A is the better at prediction, because he had a smaller sum of deviations, 1, than interviewer B, with a sum of 14. The following table shows these results.

Predicted Minus Observed for Two Interviewers				
		Interviewer		
Observed	Mr. A	Ms . B	Mr. A	Ms. B
Y_i	Y_i'	Y_i'	$Y_i - Y_i'$	$Y_i - Y_i'$
23	38	21	-15	2
18	34	15	-16	3
35	16	32	19	3
10	10	8	0	2
27	14	23	13	4
			1	14

This goes against common sense. In this case large positive deviations cancel out large negative deviations, leaving what appears as an almost perfect prediction for interviewer A. But that is obviously not the case.

In order to avoid this problem, it is possible to ignore the signs of the differences and then sum, that is, take the sum of the **absolute values.** For mathematical reasons, however, statisticians eliminate the signs by squaring the differences. In the example, this procedure would yield the results shown in this table:

Predicted Minus Observed Squared for Two Interviewers						
			Interviewer			
Observed	Mr. A	Ms . B	Mr. A	Ms. B	Mr. A	Ms. B
Y_i	Y_i'	Y_i'	$Y_i - Y_i'$	$Y_i - Y_i'$	$(Y_i - Y_i')^2$	$(Y_i - Y_i')^2$
23	38	21	-15	2	225	4
18	34	15	-16	3	256	9
35	16	32	19	3	361	9
10	10	8	0	2	0	4
27	14	23	13	4	169	16
			1	14	1011	42

Summing the squared differences, or squared residuals, yields the desired measure of goodness-of-fit. In this case, the smaller the number, the closer the predicted to the observed values. This is expressed in the following mathematical equation:

$$\sum_{i=1}^{N} (Y_i - Y_i')^2$$

The prediction that minimizes this sum is said to meet the **least-squares criterion.** In the example, interviewer B meets this criterion in a comparison between the two interviewers (her score of 42 is much less than interviewer A's 1011), and she would be promoted. Interviewer A would receive a pink slip.

14-4 The Regression Model

The situation using the regression model is analogous to that of the interviewers, except instead of using interviewers, predictions are made by performing a linear transformation of the **predictor variable.** The predicted value would be obtained by a linear transformation of the score. The prediction takes the form Y' = a + bX, where a and b are parameters in the regression model.

$$Y' = a + bX$$

The regression model.

In the previous example, suppose that rather than being interviewed, each applicant took a form-board test (see the following figure). A form-board is a board with holes cut out in various shapes: square, round, triangular, and so on. The goal is to put the right pegs in the right holes as fast as possible. The saying "square peg in a round hole" came from this test, which has been around for a long time.

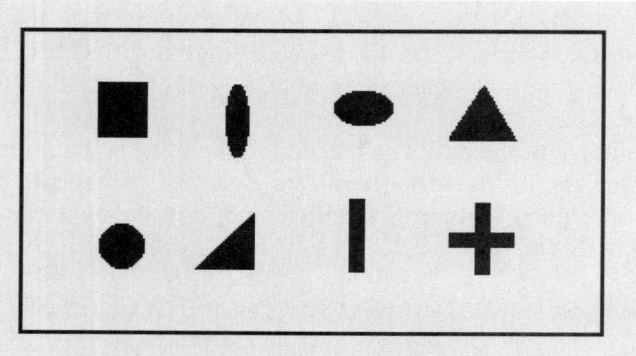

Form-board test.

The score for the test is the number of seconds it takes to complete putting all the pegs in the right holes. The data collected are shown in this table:

Example Data	
Form-Board Test	Widgets/hr
X_i	Y_i
13	23
20	18
10	35
33	10
15	27

Because the two parameters of the regression model, a and b, can take on any real value, there are an infinite number of possible models, analogous to having an infinite number of possible interviewers. The goal of regression is to select the parameters of the model so that the least-squares criterion is met, or, in other words, to minimize the sum of the squared deviations.

The procedure discussed in the last chapter, that of transforming the scale of X to the scale of Y, such that both have the same mean and standard deviation, will not work in this case, because of the prediction goal.

A number of possible models will now be examined where

X_i is the number of seconds to complete the form-board task

Y_i is the number of widgets made per hour two months later

Y'_i is the predicted number of widgets

For the first model, let a = 10 and b = 1, attempting to predict the first score perfectly. In this case the regression model becomes

$$Y' = a + bX = 10 + 1 * X$$

The first score ($X_1 = 13$) would be transformed into a predicted score of $Y_1' = 10 + (1 * 13) = 23$. The second predicted score, where $X_2 = 20$ would be $Y_2' = 10 + (1 * 20) = 30$. The same procedure is then applied to the last three scores, resulting in predictions of 20, 43, and 25, respectively, as shown in the Predicted column of this table:

Computing the Sum of Squared Residuals Where a = 10 and b = 1				
Form-Board X_i	Widgets/hr Y_i	Predicted $Y'_i = a + bX_i$	Residuals $(Y_i - Y'_i)$	Squared Residuals $(Y_i - Y'_i)^2$
13	23	23	0	0
20	18	30	-12	144
10	35	20	15	225
33	10	43	-33	1089
15	27	25	2	4
				1462

Comparing these predicted scores to the actual number of widgets of made (the second column), it can be seen that the model does a good job of prediction for the first and last applicant, but the middle applicants are poorly predicted. The sum of squared residuals in this case is 1462. Because we want the model work for all applicants, some other values for the parameters must be tried. The selection of the parameters for the second model is based on the observation that the longer it takes to put the form board together, the fewer the number of widgets made. When the tendency is for one variable to increase while the other decreases, the relationship between the variables is said to be *inverse*. In order to model an **inverse relationship,** a negative value of b must be used in the regression model. Let's try parameters of a = 36 and b = -1. This table shows the results:

Computing the Sum of Squared Residuals Where a = 36 and b = -1				
Form-Board X_i	Widgets/hr Y_i	Predicted $Y'_i = a + bX_i$	Residuals $(Y_i - Y'_i)$	Squared Residuals $(Y_i - Y'_i)^2$
13	23	23	0	0
20	18	16	2	4
10	35	26	9	81
33	10	3	7	49
15	27	21	6	36
				170

This model, with a sum of squared residuals equal to 170, fits the data much better than did the first model. Fairly large deviations are noted for the third applicant, which might be reduced by increasing the value of the additive component of the transformation, a. Thus a model with a = 41 and b = -1 is tried, and the results are in the following table:

Computing the Sum of Squared Residuals Where a = 41 and b = -1				
Form-Board X_i	Widgets/hr Y_i	Predicted $Y'_i = a + bX_i$	Residuals $(Y_i - Y'_i)$	Squared Residuals $(Y_i - Y'_i)^2$
13	23	28	-5	25
20	18	21	-3	19
10	35	31	4	16
33	10	8	2	4
15	27	26	1	1
				55

This makes the predicted values closer to the observed values on the whole, as measured by the sum of squared residuals—55 in this case. Perhaps a decrease in the value of b would make the predictions better. Hence a model where a = 32 and b = -.5 are tried. You can see the results in this table:

Computing the Sum of Squared Residuals where a = 32 and b = -.5				
Form-Board X_i	Widgets/hr Y_i	Predicted $Y'_i = a + bX_i$	Residuals $(Y_i - Y'_i)$	Squared Residuals $(Y_i - Y'_i)^2$
13	23	25.5	-2.5	6.25
20	18	22	-4	16
10	35	27	8	64
33	10	17.5	-7.5	56.25
15	27	24.5	3.5	12.25
				142.5

Since this attempt increased the sum of the squared residuals to 142.5, it obviously was not a good idea.

The point is soon reached when the question, "When do we know when to stop?" must be asked. Using this procedure, the answer must necessarily be "Never," because it is always possible to change the values of the two parameters slightly and obtain a better estimate, one that makes the sum of squared deviations smaller. The following table summarizes what is known about the problem thus far.

Sum of Squared Residuals for Various Prediction Models		
a	b	$(Y_i - Y'_i)^2$
10	1	1462
36	-1	170
41	-1	55
32	-.5	142.5

With four attempts at selecting parameters for a model, it appears that when a = 41 and b = -1, the best fitting (smallest sum of squared deviations) is found to this point in time. If the same search procedure were going to be continued, perhaps the value of a could be adjusted when b = -2 and b = -1.5, and so forth. The following program provides scroll bars to allow the student to adjust the values of a and b and view the resulting table of squared residuals. Unless the **sum of**

squared deviations is equal to zero, which is seldom possible in the real world, we will never know if it is the best possible model. Rather than throwing their hands up in despair, applied statisticians approached the mathematician with the problem and asked if a mathematical solution could be found. This is the topic of the next section—if you are willing to simply "believe," it may be skimmed without any great loss to your ability to "do" a linear regression problem.

14-5 Solving for Parameter Values That Satisfy the Least-Square Criterion

The problem is presented to the mathematician as follows: The values of a and b in the linear model $Y'_i = a + bX_i$ are to be found that minimize the algebraic expression

$$\sum_{i=1}^{N} (Y_i - Y'_i)^2$$

The mathematician begins as this figure shows:

$\sum (Y_i - Y'_i)^2$ is the expression to be minimized

$\sum (Y_i - (a + bX_i))^2$ substituting $a + bx$ for Y'

$\sum (Y_i - a - bX_i)^2$ deleting the innermost parentheses

$\sum (Y^2 + a^2 + b^2x^2 - 2aY - 2bXY + 2abX)$ squaring the expression

$\sum Y^2 + \sum a^2 + \sum b^2x^2 - 2\sum aY - 2\sum bXY + 2\sum abX$ taking the summation sign inside

Solving for the values of the slope and intercept that minimize the sum of squared residuals.

Now comes the hard part, and it requires knowledge of calculus. At this point even the mathematically sophisticated student will be asked to "believe." What the mathematician does is take the first-order partial derivative of the last form of the preceding expression with respect to b, set it equal to zero, and solve for the value of b. This is the method that mathematicians use to solve for minimum and maximum values. Completing this task, the solution for the slope of the regression model becomes:

$$b = \frac{N\sum XY - \sum X \sum Y}{N\sum X^2 - (\sum X)^2}$$

Using a similar procedure to find the value of a yields the least-square solution for the intercept:

$$a = \bar{Y} - b\bar{X}$$

The optimal values for a and b can be found by doing the appropriate summations, plugging them into the equations, and solving for the results. The appropriate summations are presented here:

Finding the Sums to Compute Optimal Regression Parameters			
X_i	Y_i	X_i^2	$X_i Y_i$
13	23	169	299
20	18	400	360
10	35	100	350
33	10	1089	330
15	27	225	405
SUM 91	113	1983	1744

The sums in the table may be written in the shorthand mathematical summation notation as follows:

$$N = 5$$

$$\sum X = 91$$

$$\sum Y = 113$$

$$\sum X^2 = 1983$$

$$\sum XY = 1744$$

Intermediate sums necessary to solve for the optimal slope and intercept.

The result of these calculations (see the following figure) is a regression model of the form.

$$b = \frac{N\sum XY - \sum X \sum Y}{N\sum X^2 - \left(\sum X\right)^2}$$

$$b = \frac{5*1744 - (91*113)}{5*1983 - 91^2}$$

$$b = \frac{8720 - 10283}{9915 - 8281}$$

$$b = \frac{-1563}{1634} = -.9565$$

Solving the example for an optimal solution for the slope.

Solving for the a parameter, after solving for the b parameter, is somewhat easier, as the following figure shows:

$$\overline{X} = 18.2$$
$$\overline{Y} = 22.6$$
$$b = -.957$$
$$a = \overline{Y} - b\overline{X}$$
$$a = 22.6 - (-.957 * 18.2)$$
$$a = 40.01$$

The solution for the optimal value of the intercept for the example data.

This procedure results in an optimal model. That is, no other values of a and b will yield a smaller sum of squared deviations. (The mathematician is willing to bet the family farm on this result.) A demonstration of this fact will be done for this problem shortly.

In any case, both the number of pairs of numbers (five) and the integer nature of the numbers made this problem "easy." This "easy" problem resulted in considerable computational effort. Imagine what a "difficult" problem with hundreds of pairs of decimal numbers would be like. That is why a bivariate, or two-variable, statistics mode is available on many calculators.

14-6 Using Statistical Calculators to Solve for Regression Parameters

Most statistical calculators require a number of steps to solve regression problems (the specific keystrokes required for the steps vary for the different makes and models of calculators; please consult your calculator manual for details):

1. Put the calculator in bivariate statistics mode. (This step is not necessary on some calculators.)

2. Clear the statistical registers.

3. Enter the pairs of numbers. (Some calculators verify the number of numbers entered at any point in time on the display.)

4. Find the values of various statistics including
   The mean and standard deviation of both X and Y
   The correlation coefficient (r)
   The parameter estimates of the regression model
   The slope (b)
   The intercept (a)

The results of these calculations for the example problem are shown in the following figure:

$$\overline{X} = 18.2 \qquad s_X = 9.04$$
$$\overline{Y} = 22.6 \qquad s_Y = 9.40$$
$$r = -0.92$$
$$a = 40.01$$
$$b = -.957$$

Summary of statistics in a regression problem.

The discussion of the correlation coefficient is left for the next chapter. All that is important at the present time is the ability to calculate the value in the process of performing a **regression analysis.** The value of the correlation coefficient will be used in a later formula in this chapter. Here's a summary of the regression statistics we've been working with:

$$\bar{X} = 18.2$$
$$\bar{Y} = 22.6$$
$$b = -.957$$
$$a = \bar{Y} - b\bar{X}$$
$$a = 22.6 - (-.957 * 18.2)$$
$$a = 40.01$$

Summary of regression statistics.

14-7 Demonstration of Optimal Parameter Estimates

Using either the algebraic expressions developed by the mathematician or the calculator, the **optimal regression model** results in the following:

$$Y'_i = 40.01 - .957X_i$$

Regression model resulting from the example data.

Applying procedures identical to those used on earlier **nonoptimal regression models,** the residuals (deviations of observed and predicted values) are found, squared, and summed to find the sum of squared deviations.

Computing the Sum of Squared Residuals where a = 40.01 and b = -.9565				
Form-Board X_i	Widgets/hr Y_i	Predicted $Y'_i = a + bX_i$	Residuals $(Y_i - Y'_i)$	Squared Residuals $(Y_i - Y'_i)^2$
13	23	27.58	-4.57	20.88
20	18	20.88	-2.78	8.28
10	35	30.44	4.55	20.76
33	10	8.44	1.56	2.42
15	27	25.66	1.34	1.80
				54.14

Note that the sum of squared deviations $((Y_i - Y'_i)^2 = 54.14)$ is smaller than the previous low of 55.0, but not by much. The mathematician is willing to guarantee that this is the smallest sum of squared deviations that can be obtained by using any possible values for a and b.

The bottom line is that the equation $Y'_i = 40.01 - .957X_i$ will be used to predict the number of widgets per hour that potential employees will make, given the scores that they made on the form-board test. The prediction will not be perfect, but it will be the best available, given the data and the form of the model.

14-8 Scatterplots and the Regression Line

The preceding has been an algebraic presentation of the logic underlying the regression procedure. There is a one-to-one correspondence between algebra and geometry, and some students have an easier time understanding a visual presentation of an algebraic procedure, so let's look at one now. The data will be represented as points on a scatterplot, while the regression equation will be represented by a straight line, called the **regression line.**

A **scatterplot** or scattergram is a visual representation of the relationship between the X and Y variables. First, the X and Y axes are drawn with equally spaced markings to include all values of that variable that occur in the sample. In the example problem, X, the seconds to put the form-board together, would have to range between 10 and 33, the lowest and highest values that occur in the sample. A similar value for the Y variable, the number of widgets made per hour, is from 10 to 35. If the axes do not start at zero, as in the present case where they both start at 10, a small space is left before the line markings to indicate this fact.

The paired or **bivariate data** (two variables, X and Y) are represented as **vectors** or points on this graph. The point is plotted by finding the intersection of the X and Y scores for that pair of values. For example, the first point would be located at the intersection of and X = 13 and Y = 23. The first point and the remaining four points are presented on the following graph:

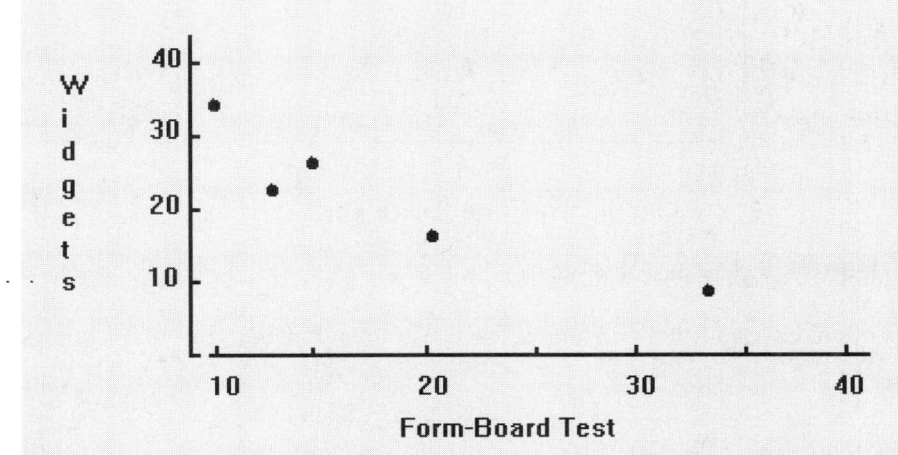

The regression line is drawn by plotting the X and Y' values. The next figure presents the five X and Y' values that were found on the regression table of observed and predicted values. Note that the first point would be plotted as (13, 27.57) the second point as (20, 20.88), and so on.

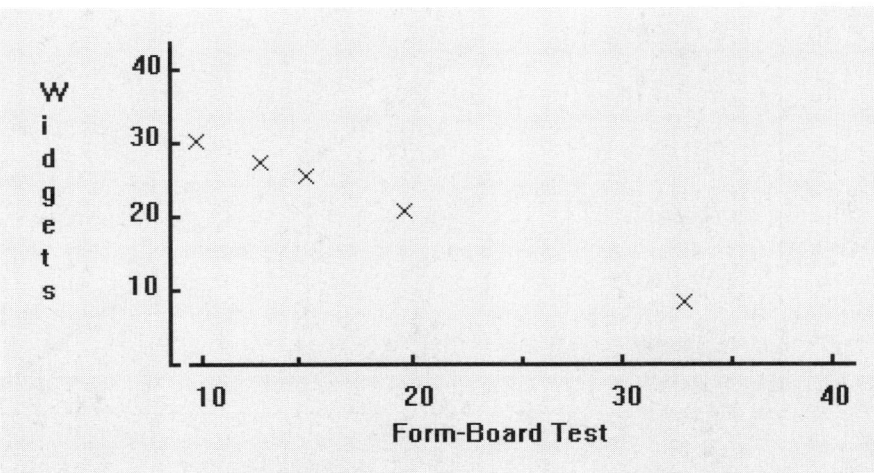

Drawing the regression line on a scatterplot by plotting values of X and Y'.

Note that all the points fall on a straight line. If every possible Y' were plotted for every possible X, then a straight line would be formed. The equation $Y' = a + bX$ defines a straight line in a two dimensional space. The easiest way to draw the line is to plot the two extreme points (the points corresponding to the smallest and largest X), and connect these points with a straightedge. Any two points would actually work, but the two extreme points give a line with the least drawing error. The a value is sometimes called the intercept and defines where the line crosses the Y-axis. This does not happen very often in actual drawings, because the axes do not begin at zero, that is, there is a break in the line. The following illustrates how to draw the regression line.

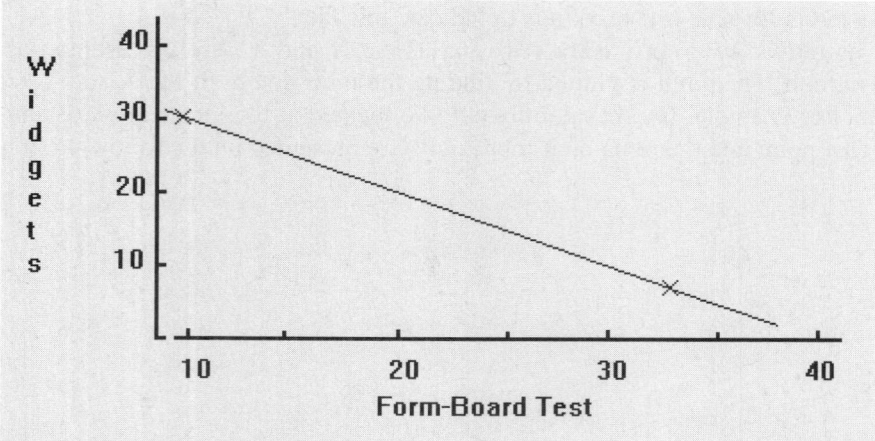

Drawing a regression line by plotting two distant points.

Most often the scatterplot and regression line are combined, as this graph shows:

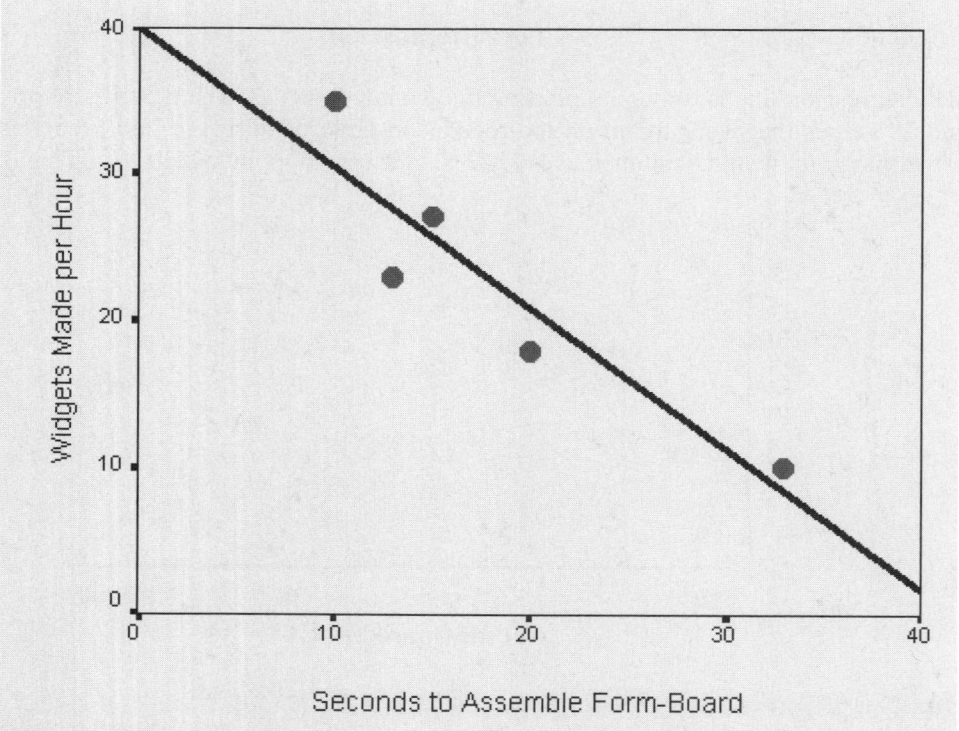

Scatterplot with regression line.

14-9 The Standard Error of Estimate

The standard error of estimate is a measure of error in prediction. It is symbolized as $s_{Y.X}$, read as s sub Y dot X. The notation is used to mean the standard deviation of Y given the value of X is known. The standard error of estimate is defined by the formula

$$s_{YX} = \sqrt{\frac{\sum_{i-1}^{N} (Y_i - Y'_i)^2}{N - 2}}$$

As such it may be thought of as the average deviation of the predicted from the observed values of Y, except the denominator is not N, but $N - 2$, the degrees of freedom for the regression procedure. One degree of freedom is lost for each of the parameters estimated, a and b. Note that the numerator is the same as in the least-squares criterion.

$$s_{YX} = \sqrt{\frac{\sum_{i-1}^{N} (Y_i - Y'_i)^2}{N - 2}}$$

$$s_{Y} = \sqrt{\frac{\sum_{i-1}^{N} (Y_i - \bar{Y})^2}{N - 1}}$$

Comparing the definitional formulas for the standard error of estimate and the standard deviation of Y.

The **standard error of estimate** is a standard deviation type of measure. Note the similarity of the definitional formula of the standard deviation of Y to the definitional formula for the standard error of measurement.

Two differences appear. First, the standard error of estimate divides the sum of squared deviations by $N - 2$, rather than $N - 1$. Second, the standard error of estimate finds the sum of squared differences around a predicted value of Y, rather than the mean.

The similarity of the two measures may be resolved if the standard deviation of Y is conceptualized as the error around a predicted Y of $Y'_i = a$. When the least-squares criterion is applied to this model, the optimal value of a is the mean of Y. In this case only one degree of freedom is lost because only one parameter is estimated for the regression model.

The standard error of estimate may be calculated from the definitional formula previously given. The computation is difficult, however, because the entire table of differences and squared differences must be calculated. Because the numerator has already been found, the calculation for the example data is relatively easy, as the following formula shows:

$$s_{YX} = \sqrt{\frac{\sum_{i=1}^{N} (Y_i - Y'_i)^2}{N - 2}}$$

$$s_{YX} = \sqrt{\frac{54.14}{5-2}} = 4.248$$

Example computation using the definitional formula for the standard error of estimate.

The calculation of the standard error of estimate is simplified by the following formula, called the **computational formula for the standard error of estimate.** The computation is easier because the statistical calculator computes the correlation coefficient when finding a regression line. The computational formula for the standard error of estimate will always give the same result, within rounding error, as the definitional formula. The computational formula may look more complicated, but it does not require the computation of the entire table of differences between observed and predicted Y scores. The computational formula is

$$s_{YX} = \sqrt{\frac{N-1}{N-2} s_y^2 (1 - r^2)}$$

where:
s_y^2 *is the variance of Y*
r^2 *is the correlation coefficient squared*

The computational formula for the standard error of estimate is most easily and accurately computed by temporarily storing the values for s_Y^2 and r^2 in the calculator's memory and recalling them when needed. Using this formula to calculate the standard error of estimate with the example data produces the following results:

$$s_{YX} = \sqrt{\frac{N-1}{N-2} s_y^2 (1 - r^2)}$$

$$s_{YX} = \sqrt{\frac{5-1}{5-2} 9.40^2 (1 - (-.92)^2)}$$

$$s_{YX} = \sqrt{18.06} = 4.25$$

Example of the application of the computational formula for the standard error of estimate.

Note that the result is the same as the result from the application of the definitional formula, within rounding error.

The standard error of estimate is a measure of error in prediction. The larger its value, the less well the regression model fits the data, and the worse the prediction.

14-10 Conditional Distributions

A **conditional distribution** is a distribution of a variable given a particular value of another variable. For example, a conditional distribution of number of widgets made exists for each possible value of number of seconds to complete the form-board test. Conceptually, suppose that an infinite number of applicants had made the same score of 18 on the **form-board test.** If everyone were hired, not everyone would make the same number of widgets three months later. The distribution of scores that results would be called the conditional distribution of Y (widgets) given X (form-board test score). The relationship between X and Y in this case is often symbolized by Y|X. The conditional distribution of Y given that X was 18 would be symbolized as Y|X = 18.

It is possible to model the conditional distribution with the normal curve. In order to create a normal curve model, it is necessary to estimate the values of the parameters of the model, $\mu_{Y|X}$ and $\sigma_{Y|X}$. The best estimate of $\mu_{Y|X}$ is the predicted value of Y, Y', given X equals a particular value. This is found by entering the appropriate value of X in the regression equation, Y' = a + bX. In the example, the estimate of $\mu_{Y|X}$ for the conditional distribution of number of widgets made given X = 18, would be Y' = 40.01 − .957 * 18 = 22.78. This value is also called a **point estimate,** because it is the best guess of Y when X is a given value.

The standard error of estimate is often used as an estimate of $\sigma_{Y|X}$ for all the conditional distributions. This assumes that all conditional distributions have the same value for this parameter. One interpretation of the standard error of estimate, then, is an estimate of the value of $\sigma_{Y|X}$ for all possible conditional distributions or values of X. The conditional distribution which results when X = 18 is presented here.

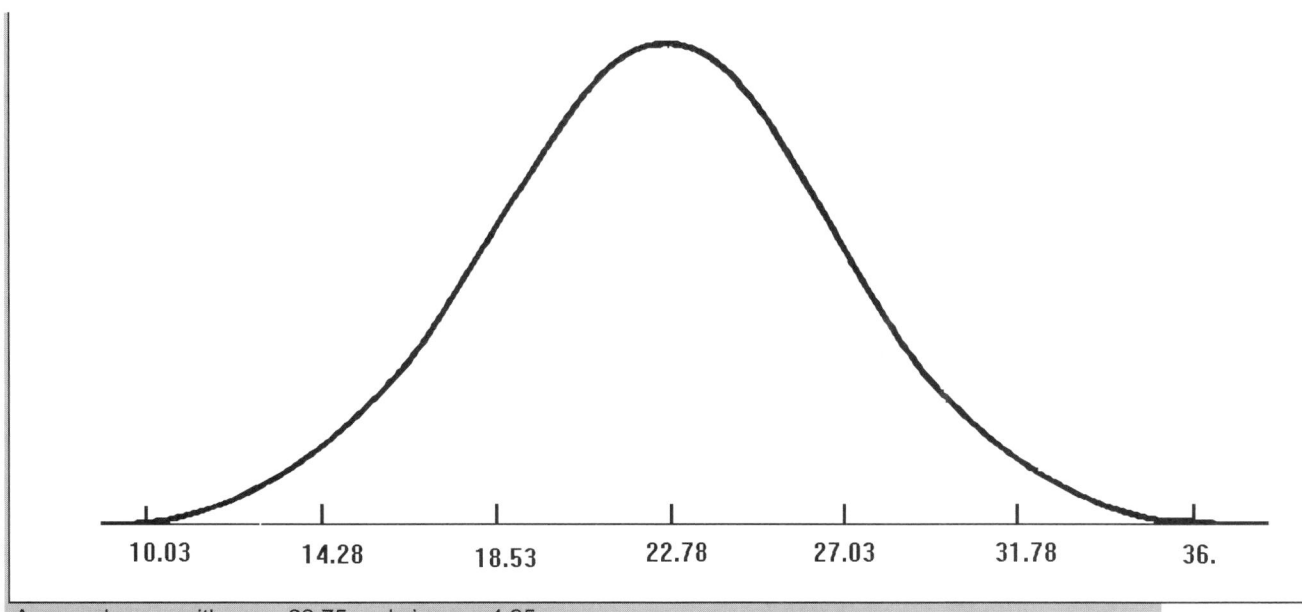

| 10.03 | 14.28 | 18.53 | 22.78 | 27.03 | 31.78 | 36. |

A normal curve with mu = 22.75 and sigma = 4.25.

It is somewhat difficult to visualize all possible conditional distributions in only two dimensions, although the following illustration attempts the relatively impossible. If a hill can be visualized with the middle being the regression line, the vision would be essentially correct.

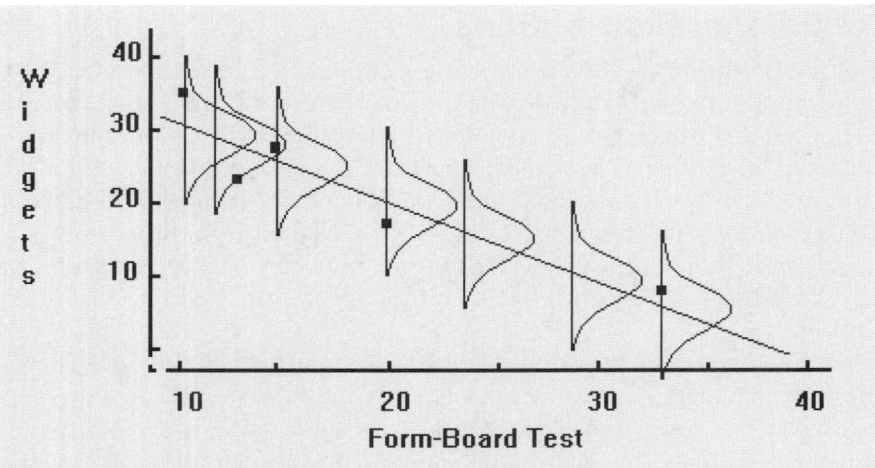

Representation of conditional distributions on a scatterplot.

The conditional distribution is a model of the distribution of points around the regression line for a given value of X. The conditional distribution is important in this text mainly for the role it plays in computing an interval estimate.

14-11 Interval Estimates

The error in prediction may be incorporated into the information given to the client by using interval estimates rather than point estimates. A point estimate is the predicted value of Y, Y'. While it gives the best possible prediction, as defined by the least-squares criterion, the prediction is not perfect. The **interval estimate** presents two values; low and high, between which some percentage of the observed scores are likely to fall. For example, if a person applying for a position manufacturing widgets made a score of X = 18 on the form-board test, a point estimate of 22.78 would result from the application of the regression model and a ninety-five percent confidence interval estimate might be from 14.25 to 31.11. Using a ninety-five percent **confidence interval,** it can be said that 95 times out of 100 the number of widgets made per hour by an applicant making a score of 18 on the form-board test would be between 14.25 and 31.11.

The concept of the conditional distribution is critical to understanding the assumptions made when calculating an interval estimate. If the conditional distribution for a value of X is known, then finding an interval estimate is reduced to a problem that has already been solved in an earlier chapter. That is, what two scores on a normal distribution with parameters cut off some middle percent of the distribution? While any percentage could be found, the standard value is a 95% confidence interval.

For example, the parameter estimates of the conditional distribution of X = 18 are $\mu_{Y|X} = 22.78$ and $\sigma_{Y|X} = 4.25$. The two scores which cut off the middle 95% of that distribution are 14.25 and 31.11. Use the Probability Calculator to find the middle area. Here are the steps, as shown in the following figure:

1. Select Confidence Interval.

2. Click on Normal Curve.

3. Enter Mu, Sigma, and Probability values.

4. Click the left-arrow button.

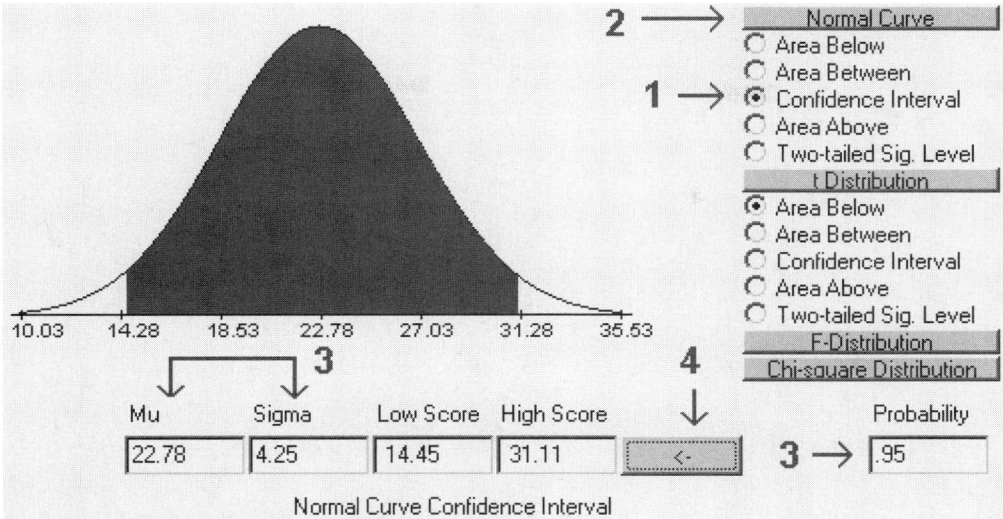

Finding a confidence interval using the Probability Calculator.

Other sizes of confidence intervals could be computed by changing the value of the probability in the Probability Calculator.

Interpretation of the confidence interval for a given score of X necessitates several assumptions. First, the conditional distribution for that X is a normal distribution. Second, $\mu_{Y|X}$ is correctly estimated by Y', that is, the relationship between X and Y can be adequately modeled by a straight line. Third, $\sigma_{Y|X}$ is correctly estimated by $s_{Y.X}$, which means assuming that all conditional distributions have the same estimate for $\sigma_{Y|X}$.

14-12 Regression Analysis Using SPSS

To use the Regression command in SPSS, select Analyze/Regression/Linear, as the following figure shows:

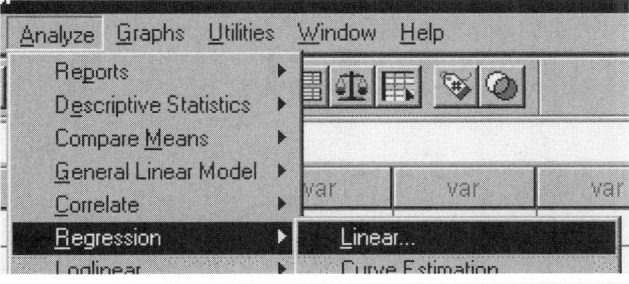

Computing the regression model using SPSS.

On the screen that follows, click the Save button. In the pop-up Linear Regression:Save dialog box, check Unstandardized under both Predicted Values and Residuals, as shown in the following figure. Click Continue and then OK, and the program produces a simple **linear regression** and creates two new variables in the data editor: one with the predicted values of Y and the other with the residuals.

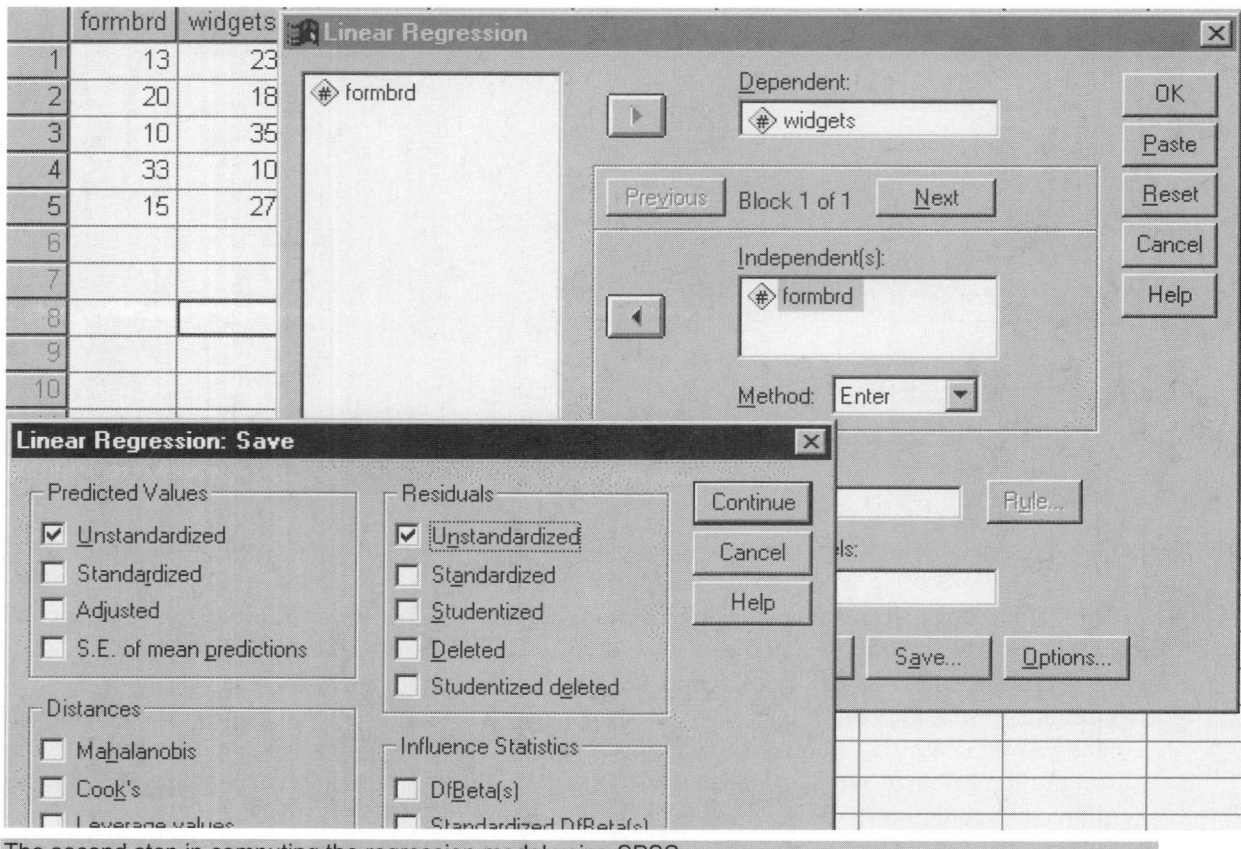

The second step in computing the regression model using SPSS.

The first table of the output (see the following figure) includes a standard error of estimate.

Model Summary[b]

Model	R	R Square	Adjusted R Square	Std. Error of the Estimate
1	.920[a]	.847	.795	4.25

a. Predictors: (Constant), FORMBRD

b. Dependent Variable: WIDGETS

SPSS model summary table for regression procedure.

The **regression coefficients,** or values for a (intercept) and b (slope), are found in the third table in the SPSS output. The B values under Unstandardized Coefficients are the intercept (Constant) and slope (FORMBRD). The Beta values under Standardized Coefficients correspond to the regression weights when both x and y have been converted to z-scores. In this case, z of y is predicted from z of x.

Coefficients[a]

	Model	Unstandardized Coefficients		Standardized Coefficients		
		B	Std. Error	Beta	t	Sig.
1	(Constant)	40.009	4.682		8.546	.003
	FORMBRD	-.957	.235	-.920	-4.069	.027

a. Dependent Variable: WIDGETS

The third table containing the output from the SPSS Regression procedure.

The optional Save command generates two new variables in the data file: pre_1 (predicted Y) and res_1 (residuals). The data file will look like the following figure:

	formbrd	widgets	pre_1	res_1
1	13	23	27.57405	-4.57405
2	20	18	20.87821	-2.87821
3	10	35	30.44370	4.55630
4	33	10	8.44308	1.55692
5	15	27	25.66095	1.33905

The addition of the pre_1 and res_1 columns to the data file after using the Regression command in SPSS.

Summary

Regression models are powerful tools for predicting a score based on some other score. They involve a linear transformation of the predictor variable into the predicted variable. The parameters of the linear transformation are selected such that the least-squares criterion is met, resulting in an optimal model. The model can then be used in the future to predict either exact scores, called point estimates, or intervals of scores, called interval estimates.

Graphically the relationship between two variables can be represented in a scatterplot. A point on the scatterplot represents each pair of scores. The regression line is represented as a line on the graph. The closer the points fall to the line, the better the fit of the model to the data.

Chapter

15

Correlation

Key Terms

absolute measure
causal variable
causation
correlation
correlation coefficient

correlation matrix
invariant
negative correlation coefficient
outlier
path analysis

positive correlation coefficient
regression
relative measure
squared correlation coefficient

The Pearson Product-Moment Correlation Coefficient (r), or **correlation coefficient** for short, is a measure of the degree of linear relationship between two variables, usually labeled X and Y. While in **regression** the emphasis is on predicting one variable from the other, in **correlation** the emphasis is on the degree to which a linear model may describe the relationship between two variables. In regression the interest is directional—one variable is predicted and the other is the predictor; in correlation the interest is nondirectional—the relationship is the critical aspect.

The computation of the correlation coefficient is most easily accomplished with the aid of a statistical calculator. The value of r was found on a statistical calculator during the estimation of regression parameters in the last chapter. Although definitional formulas will be given later in this chapter, you are encouraged to review the procedure for obtaining the correlation coefficient on the calculator at this time.

The correlation coefficient may take on any value from plus one to minus one:

$$-1.00 \le r \le +1.00$$

The sign of the correlation coefficient (+ or −) defines the direction of the relationship, either positive or negative. A **positive correlation coefficient** means that as the value of one variable increases, the value of the other variable increases; as one decreases, the other decreases. A **negative correlation coefficient** indicates that as one variable increases, the other decreases, and vice-versa.

Taking the absolute value of the correlation coefficient measures the strength of the relationship. A correlation coefficient of $r = .50$ indicates a stronger degree of linear relationship than one of $r = .40$. Likewise a correlation coefficient of $r = -.50$ shows a greater degree of relationship than one of $r = .40$. Thus a correlation coefficient of zero ($r = 0.0$) indicates the absence of a linear relationship and correlation coefficients of $r = +1.0$ and $r = -1.0$ indicate a perfect linear relationship.

15-1 Understanding and Interpreting the Correlation Coefficient

The correlation coefficient may be understood by various means, each of which will be examined in turn.

15-1a Scatterplots

The scatterplots can perhaps best illustrate how the correlation coefficient changes as the linear relationship between the two variables is altered. When $r = 0.0$, the points scatter widely about the plot, the majority falling roughly in the shape of a circle. As the linear relationship increases, the circle becomes more and more elliptical in shape until the limiting case is reached ($r = 1.00$ or $r = -1.00$) and all the points fall in a straight line.

A number of scatterplots and their associated correlation coefficients are presented here so that you may better estimate the value of the correlation coefficient based on a scatterplot.

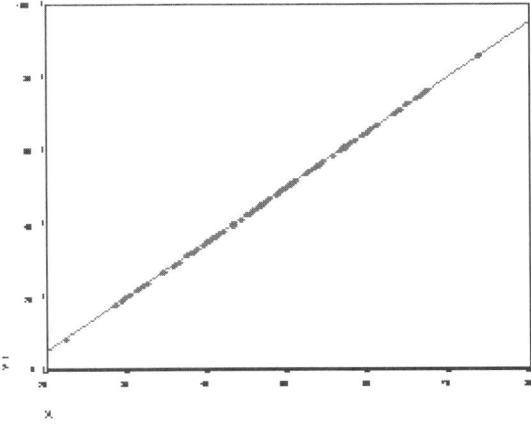

r = +1.00.

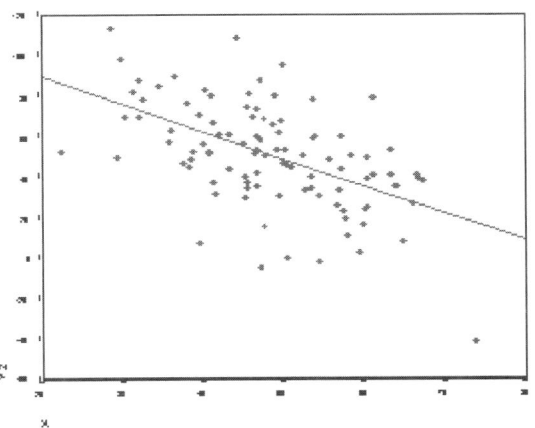

r = -.54.

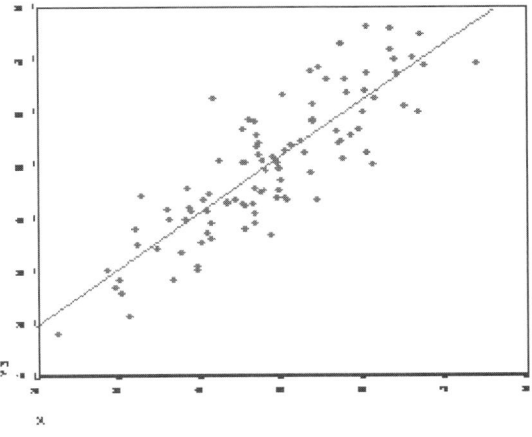

r = +.85.

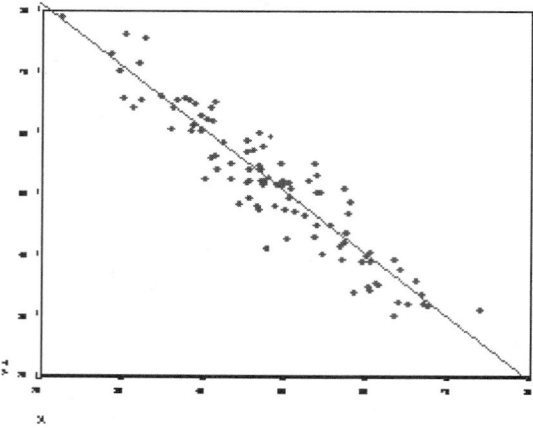

r = -.94.

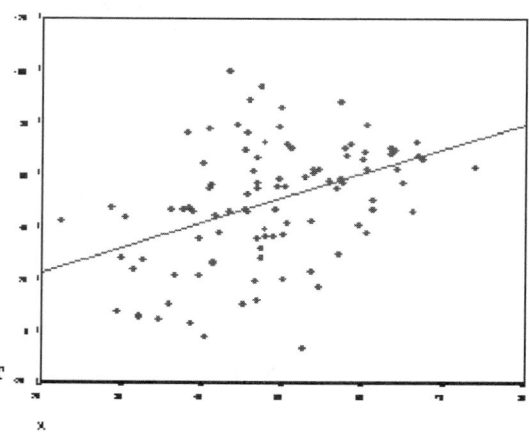

r = +.42.

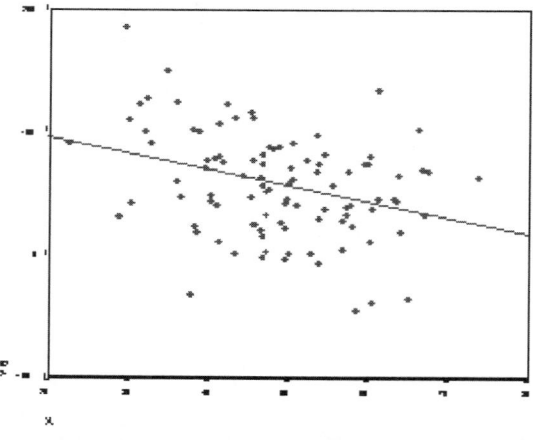

r = -.33.

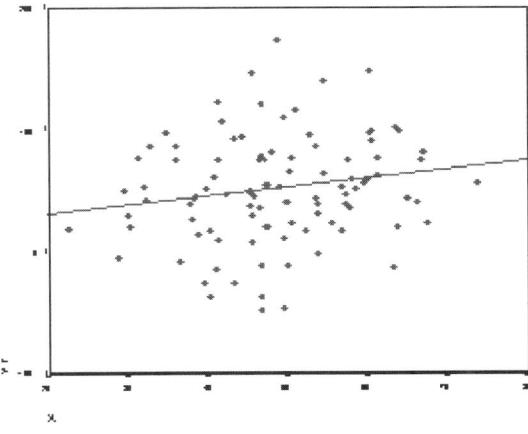

r = +.17.

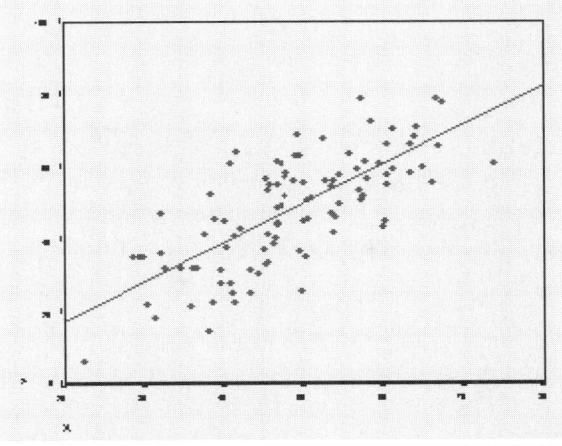

r = +.39

15-1b Slope of the Regression Line of Z-Scores

The correlation coefficient is the slope (b) of the regression line when both the X and Y variables have been converted to z-scores. The larger the size of the correlation coefficient, the steeper the slope.

This interpretation of the correlation coefficient is perhaps best illustrated with an example involving numbers. The raw score values of the X and Y variables are presented in the first two columns of the following table.

	X	Y	z_X	z_Y
Transforming Raw Scores to Z-Scores				
	12	33	-1.07	-0.61
	15	31	-0.07	-1.38
	19	35	-0.20	0.15
	25	37	0.55	.92
	32	37	1.42	.92
	20.60	34.60	0.0	.0
$\overline{X}$ =	8.02	2.61	1.0	1.0

The second two columns are the X and Y columns transformed using the z-score transformation. That is, the mean is subtracted from each raw score in the X and Y columns and then the result is divided by the sample standard deviation, as these formulas show:

$$z_X = \frac{X - \bar{X}}{s_X}$$

$$z_Y = \frac{Y - \bar{Y}}{s_Y}$$

Formulas to transform from raw scores to z-scores.

There are two points to be made with these numbers:

1. The correlation coefficient is **invariant** (does not change) under a linear transformation of either X and/or Y.

2. The slope of the regression line when both X and Y have been transformed to z-scores is the correlation coefficient.

Computing the correlation coefficient first with the raw scores X and Y yields r = 0.85. Computing the correlation coefficient with z_X and z_Y yields the same value, r = 0.85. Since the z-score transformation is a special case of a linear transformation (X' = a + bX), it may be proven that the correlation coefficient is invariant under a linear transformation of either X and/or Y. You can verify this by computing the correlation coefficient using X and z_Y or Y and z_X. What this means essentially is that changing the scale of either the X or the Y variable will not change the size of the correlation coefficient, as long as the transformation conforms to the requirements of a linear transformation.

The fact that the correlation coefficient is the slope of the regression line when both X and Y have been converted to z-scores can be demonstrated by computing the regression parameters predicting z_X from z_Y or z_Y from z_X. In either case the intercept or additive component of the regression line (a) will be zero or very close, within rounding error. The slope (b) will be the same value as the correlation coefficient, again within rounding error. This relationship may be illustrated as follows:

$$z_Y' = r\, z_X$$

$$z_X' = r\, z_Y$$

The correlation coefficient is the slope of the regression line when both X and Y have been converted to z-scores.

15-1c Variance Interpretation

The **squared correlation coefficient** (r^2) is the proportion of variance in Y that can be accounted for by knowing X. Conversely, it is the proportion of variance in X that can be accounted for by knowing Y.

One of the most important properties of variance is that it may be partitioned into separate additive parts. For instance, consider the shoe size example earlier in this text.

If the scores in this distribution were partitioned into two groups, one for males and one for females, the distributions could be represented as follows:

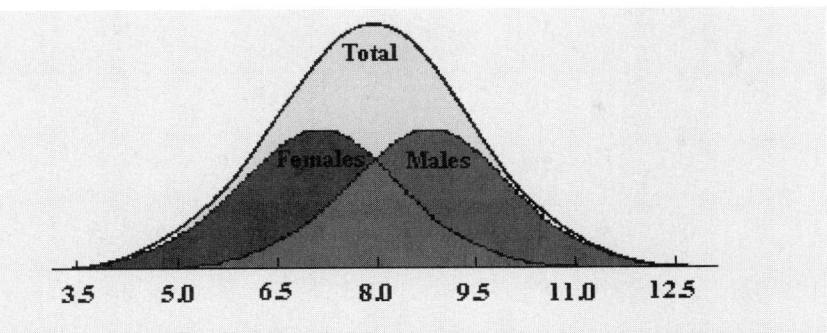

The total variance is partitioned into two parts, one for males and the other for females.

If you know the gender of an individual, you know something about that person's shoe size, because the shoe sizes of males are on the average, somewhat larger than females. The variance within each distribution, male and female, is variance that cannot be predicted on the basis of gender, or error variance, because if you know the gender of an individual, you do not know exactly what that person's shoe size is.

Rather than having just two levels, the X variable will usually have many levels. The preceding argument may be extended to encompass this situation. It can be shown that the total variance is the sum of the variance that can be predicted and the error variance, or variance that cannot be predicted. This relationship is summarized here:

$$s^2_{TOTAL} = s^2_{PREDICTED} + s^2_{ERROR}$$

or

$$s^2_{PREDICTED} = s^2_{TOTAL} - s^2_{ERROR}$$

Relationships between variance components.

The correlation coefficient squared is equal to the ratio of predicted to total variance:

$$r^2 = \frac{s^2_{PREDICTED}}{s^2_{TOTAL}}$$

The correlation coefficient squared is the proportion of predicted variance.

This formula may be rewritten in terms of the error variance (rather than the predicted variance):

$$r^2 = \frac{s^2_{TOTAL} - s^2_{ERROR}}{s^2_{TOTAL}}$$

$$r^2 = \frac{s^2_{TOTAL}}{s^2_{TOTAL}} - \frac{s^2_{ERROR}}{s^2_{TOTAL}}$$

$$r^2 = 1 - \frac{s^2_{ERROR}}{s^2_{TOTAL}}$$

The correlation coefficient squared is one minus the error variance.

The error variance, s^2_{ERROR}, is estimated by the standard error of estimate squared, $s^2_{Y.X}$, discussed in the previous chapter. The total variance (s^2_{TOTAL}) is simply the variance of Y, s^2_Y. The formula now becomes

$$r^2 = 1 - \frac{s^2_{YX}}{s^2_Y}$$

The relationship between the correlation coefficient squared, the standard error of estimate, and the variance of Y.

Solving for $s_{Y.X}$, and adding a correction factor $(N - 1) / (N - 2)$, yields the computational formula for the standard error of estimate:

$$s_{YX} = \sqrt{\frac{(N-1)}{(N-2)} s^2_Y (1 - r^2)}$$

The computational formula for the standard error of estimate.

This captures the essential relationship between the correlation coefficient, the variance of Y, and the standard error of estimate. As the standard error of estimate becomes larger relative to the total variance, the correlation coefficient becomes smaller. Thus the correlation coefficient is a function of both the standard error of estimate and the total variance of Y. The standard error of estimate is an **absolute measure** of the amount of error in prediction, while the correlation coefficient squared is a **relative measure,** relative to the total variance.

15-2 Calculation of the Correlation Coefficient

The easiest method of computing a correlation coefficient is to use a statistical calculator or computer program. Barring that, the correlation coefficient may be computed using the following formula:

$$r = \frac{\sum_{i=1}^{N} z_X z_Y}{N-1}$$

The definitional formula for the correlation coefficient.

The example data in the following table are used to demonstrate computation using this formula. Computation is rarely done in this manner and is provided as an example of the application of the definitional formula, although this formula provides little insight into the meaning of the correlation coefficient.

Computing the Correlation Coefficient with Z-Scores				
X	Y	z_X	z_Y	$z_X * z_Y$
12	33	-1.07	-0.61	0.65
15	31	-0.07	-1.38	0.97
19	35	-0.20	0.15	-0.03
25	37	0.55	.92	0.51
32	37	1.42	.92	1.31
			SUM =	3.40

Plug in the numbers, and the formula resolves as follows:

$$r = \frac{\sum_{i-1}^{N} z_X z_Y}{N-1} = \frac{3.40}{4} = .85$$

Applying the computational formula for r using example data.

15-3 The Correlation Matrix

A convenient way of summarizing a large number of correlation coefficients is to put them in a single table, called a **correlation matrix.** A correlation matrix is a table of all possible correlation coefficients between a set of variables. For example, suppose a questionnaire of the following form was given to 40 Psychology 121 students:

age: What is your age? _____

know (Number of correct answers out of 10 possible to a geography quiz that consisted of locating 10 states on a map of the United States): _____

visit: How many states have you visited? _____

comair (No = 0, Yes = 1): Have you ever flown on a commercial airliner? _____

sex (Male = 1, Female = 2): _____

The responses produced the following data matrix (only the first 7 of the 40 subjects are visible in the figure):

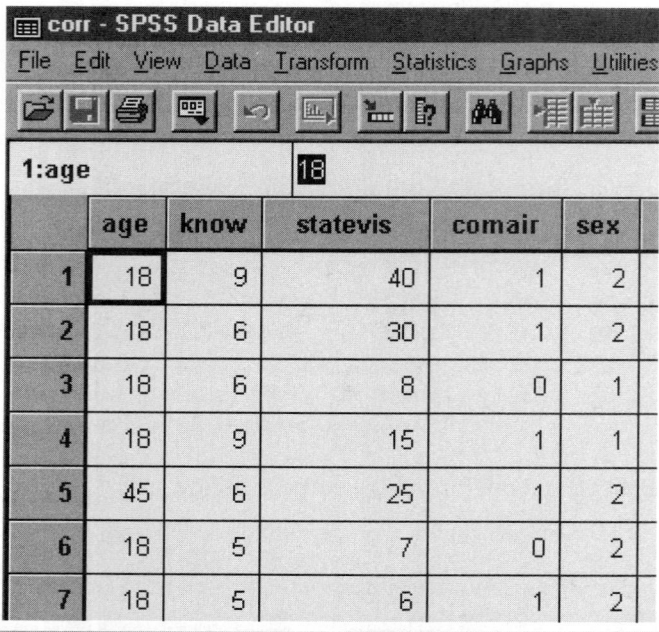

Entering the data into the data editor of SPSS.

Since there are five questions on the example questionnaire there are 5 * 5 = 25 different possible correlation coefficients to be computed. Each computed correlation is then placed in a table with variables as both rows and columns at the intersection of the row and column variable names. For example, you could calculate the correlation between age and know, age and statevis, age and comair, age and sex, know and statevis, and so on.

You would not need to calculate all possible correlation coefficients, however, because the correlation of any variable with itself is necessarily 1.00. Thus the diagonals of the matrix need not be computed. In addition, the correlation coefficient is nondirectional. That is, it doesn't make any difference whether the correlation is computed between age and know with age as X and know as Y or know as X and age as Y. For this reason the correlation matrix is symmetrical around the diagonal. In the example case then, of the 25 possible correlation coefficients, only 10 are unique and need be computed (we eliminate the 5 diagonals and the 10 that are redundant because of symmetry).

To calculate a correlation matrix using SPSS, select Analyze/Correlate/Bivariate as shown:

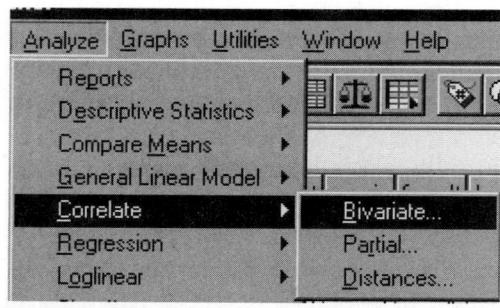

The commands to find a correlation matrix using SPSS.

Select the variables that are to be included in the correlation matrix (in this case all variables will be included), and choose the other options as shown in the following figure. Note that we are selecting means and standard deviations to be output, too.

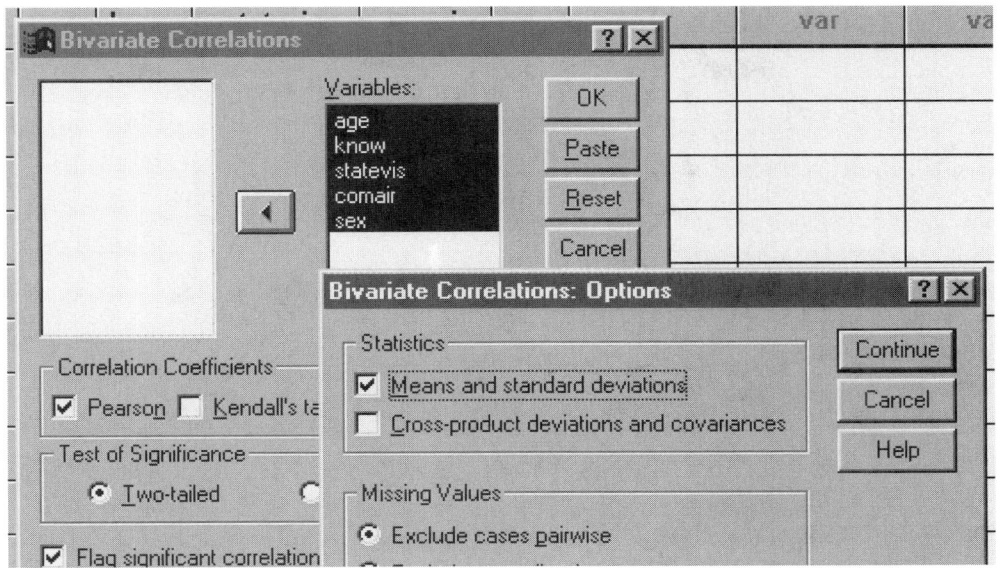

Including variables in a correlation matrix using SPSS, choosing options, and selecting optional statistics.

Click Continue and OK. The results are as follows:

Descriptive Statistics

	Mean	Std. Deviation	N
AGE	19.60	4.68	40
KNOW	6.38	2.20	40
STATEVIS	18.25	13.48	40
COMAIR	.68	.47	40
SEX	1.43	.50	40

The output from the optional statistics command in SPSS correlations.

Correlations

		AGE	KNOW	STATEVIS	COMAIR	SEX
Pearson Correlation	AGE	1.000	.027	.224	.194	.173
	KNOW	.027	1.000	.283	.169	-.078
	STATEVIS	.224	.283	1.000	.418**	-.164
	COMAIR	.194	.169	.418**	1.000	-.051
	SEX	.173	-.078	-.164	-.051	1.000
Sig. (2-tailed)	AGE	.	.867	.165	.231	.286
	KNOW	.867	.	.076	.298	.630
	STATEVIS	.165	.076	.	.007	.311
	COMAIR	.231	.298	.007	.	.753
	SEX	.286	.630	.311	.753	.
N	AGE	40	40	40	40	40
	KNOW	40	40	40	40	40
	STATEVIS	40	40	40	40	40
	COMAIR	40	40	40	40	40
	SEX	40	40	40	40	40

**. Correlation is significant at the 0.01 level (2-tailed).

The correlation matrix output using SPSS.

Interpretation of the data analysis might proceed as follows. The table of means and standard deviations indicates that the average Psychology 121 student who filled out this questionnaire was about 19 years old, could identify slightly more than six states out of ten, and had visited a little more than 18 of the 50 states. The majority (27 of the 40 students, or 68%) has flown on a commercial airplane and there were fewer females (43%) than males.

The analysis of the correlation matrix indicates that few of the observed relationships were very strong. The strongest relationship was between the number of states visited and whether or not the student had flown on a commercial airplane (r = .42) which indicates that if a student had flown, he/she was more likely to have visited more states. This is because of the positive sign on the correlation coefficient and the coding of the commercial airplane question (No = 0, Yes = 1). The positive correlation means that as X increases, so does Y: thus, students who responded that they had flown on a commercial airplane visited more states on the average than those who hadn't.

Age was positively correlated with number of states visited (r = .22) and flying on a commercial airplane (r = .19) with older students more likely both to have visited more states and flown, although the relationship was not very strong. The greater the number of states visited, the more states the student was likely to correctly identify on the map, although again relationship was weak (r = .28). Note that one of the students who said he had visited 48 of the 50 states could identify only 5 of 10 on the map.

Finally, gender of the participant was slightly correlated with both age, (r = .17) indicating that females were slightly older than males, and number of states visited (r = -.16), indicating that females visited fewer states than males These conclusions are possible because of the sign of the correlation coefficient and the way the sex variable was coded: 1 = male, 2 = female. When the correlation with sex is positive, females will have more of whatever is being measured on Y. The opposite is the case when the correlation is negative.

15-4 Cautions about Interpreting Correlation Coefficients

Correct interpretation of a correlation coefficient requires the assumption that both variables, X and Y, meet the interval property requirements of their respective measurement systems. Calculators and computers will produce a correlation coefficient regardless of whether or not the numbers are "meaningful" in a measurement sense.

As discussed in Chapter 4, "Measurement," the interval property is rarely, if ever, fully satisfied in real applications. There is some difference of opinion among statisticians about when it is appropriate to assume that the interval property is met. My personal opinion is that as long as a larger number means that the object has more of something or another, then application of the correlation coefficient is useful, although the potentially greater deviations from the interval property must be interpreted with greater caution. When the data is clearly nominal-categorical with more than two levels (such as 1 = Protestant, 2 = Catholic, 3 = Jewish, 4 = Other), interpretation of the correlation coefficient is clearly inappropriate.

An exception to the preceding rule occurs when the nominal categorical scale is dichotomous, or has two levels (1 = Male, 2 = Female). Correlation coefficients computed with data of this type on either the X and/or Y variable may be safely interpreted because the interval property is assumed to be met for these variables. Correlation coefficients computed using data of this type are sometimes given special, different names, but since they seem to add little to the understanding of the meaning of the correlation coefficient, they are not presented in this text.

15-4a Effect of Outliers

An **outlier** is a score that falls outside the range of the rest of the scores on the scatterplot. For example, if age is a variable and the sample is a statistics class, an outlier would be a retired individual. Depending upon where the outlier falls, the correlation coefficient may be increased or decreased.

An outlier that falls near where the regression line would normally fall would necessarily increase the size of the correlation coefficient, as shown in this figure:

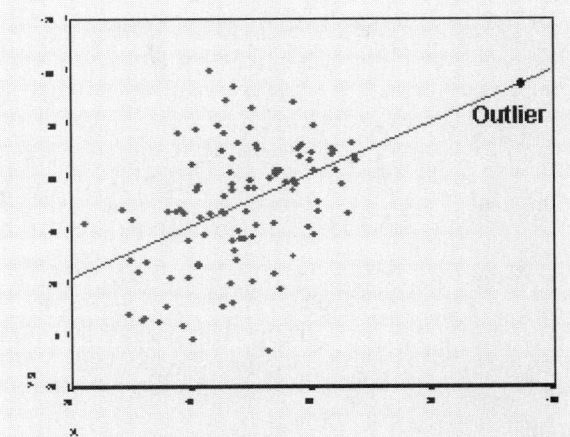

r = +.457 with outlier.

An outlier that falls some distance away from the original regression line (see the following figure) would decrease the size of the correlation coefficient.

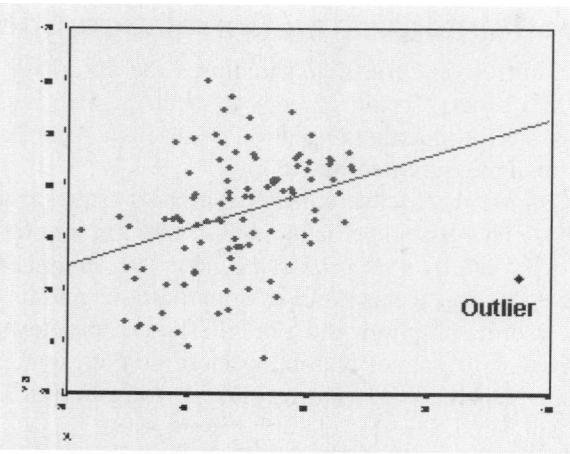

Outlier

r = +.336 with outlier.

The effect of the outliers on these examples is somewhat muted because the sample size is fairly large (N = 100). The smaller the sample size, the greater the effect of the outlier. At some point the outlier will have little or no effect on the size of the correlation coefficient.

When a researcher encounters an outlier, a decision must be made whether to include it in the data set. It may be that the respondent was deliberately malingering, giving wrong answers, or simply did not understand the question on the questionnaire. On the other hand, it may be that the outlier is real and simply different. The decision whether to include or not include an outlier remains with the researcher; he or she must justify deleting any data to the reader of a technical report, however. It is suggested that the correlation coefficient be computed and reported both with and without the outlier if there is any doubt about whether or not it is real data. In any case, the best way of spotting an outlier is by drawing the scatterplot.

15-4b Correlation and Causation

No discussion of correlation would be complete without a discussion of **causation.** It is possible for two variables to be related (correlated), but not have one variable cause another.

For example, suppose there exists a high correlation between the number of Popsicles sold and the number of drowning deaths on any day of the year. Does that mean that you should not eat Popsicles before you swim? Not necessarily. Both of the variables are related to a common variable, the heat of the day. The hotter the temperature, the more Popsicles sold and also the more people swimming, thus the more drowning deaths. This is an example of correlation without causation.

Much of the early evidence that cigarette smoking causes cancer was correlational. It may be that people who smoke are more nervous and nervous people are more susceptible to cancer. It may also be that smoking does indeed cause cancer. The cigarette companies made the former argument, while some doctors made the latter. In this case I believe the relationship is causal and therefore do not smoke.

Sociologists are very much concerned with the question of correlation and causation because much of their data is correlational. Sociologists have developed a branch of correlational analysis, called **path analysis,** precisely to determine causation from correlations. Before a correlation may imply causation, certain requirements must be met. These requirements include that the **causal variable** must temporally precede the variable it causes, and that certain relationships between the causal variable and other variables must be met.

If a high correlation was found between the age of the teacher and the students' grades, it does not necessarily mean that older teachers are more experienced, teach better, and give higher grades. Neither does it necessarily imply that older teachers are soft touches, don't care, and give higher grades. Some other explanation might also explain the results. The correlation means that older teachers give higher grades; younger teachers give lower grades. It does not explain why it is the case.

Summary

A simple correlation may be interpreted in a number of different ways: as a measure of linear relationship, as the slope of the regression line of z-scores, and as the correlation coefficient squared as the proportion of variance accounted for by knowing one of the variables. All of these interpretations are correct and, in a certain sense, mean the same thing.

A number of qualities that might affect the size of the correlation coefficient were identified. They included missing parts of the distribution, outliers, and common variables. Finally, the relationship between correlation and causation was discussed.

Chapter

16

Hypothesis Testing and Probability Theory

Key Terms

area under theoretical models
  of distributions
Bayesian Statistics
compound event
compound probabilities

conditional probability
effect
expected utility theory
hypothesis tests
null hypothesis

probability theory
sampling distribution
subjective probabilities
utility

One of the major methods of inferential statistics is hypothesis testing. Hypothesis testing is used to help minimize the number of false conclusions we draw about the nature of the world around us. Basically we want to spend our time thinking about real relationships between events rather than wasting our efforts trying to explain chance or haphazard events. Let's start with an example.

16-1 Does Caffeine Make People More Alert?

An experimental design.

Does the coffee I drink almost every morning really make me more alert? If each student drank a cup of coffee before class, would the time spent sleeping in class decrease? These questions may be answered using experimental methodology and hypothesis testing procedures.

To test the effect of caffeine on alertness in people, one experimental design would divide the classroom students into two groups; one group receiving coffee with caffeine, the other coffee without caffeine. The second group gets coffee without caffeine rather than nothing to drink because the effect of caffeine is the effect of interest, rather than the effect of ingesting liquids. The number of minutes that students sleep during that class would be recorded.

Suppose the group that got coffee with caffeine sleeps less on the average than the group that drank coffee without caffeine. On the basis of this evidence, the researcher argues that caffeine had the predicted effect.

A statistician, learning of the study, argues that such a conclusion is not warranted without performing a hypothesis test. The reasoning for this argument goes as follows: Suppose that caffeine really had no effect. Isn't it possible that the difference between the average alertness of the two groups was due to chance? That is, the individuals in the caffeine group had gotten a better night's sleep, were more interested in the class, and so forth than the members of the no-caffeine group? If the class were divided in a different manner the differences would disappear.

The purpose of the **hypothesis test** is to make a rational decision between the hypotheses of real effects and chance explanations. The scientist is never able to totally eliminate the chance explanation, but may decide that the difference between the two groups is so large that it makes the chance explanation unlikely. If this were the case, the decision would be made that the effects are real. A hypothesis test specifies how large the differences must be in order to decide that the effects are real.

The size of the difference necessary to decide that the effects are real depends on a number of factors. One factor is the amount of variability within each group, or how different the subjects within the caffeine and no-caffeine groups were from each other. If the subjects within each group were very different, then the differences between the groups would necessarily be larger in order to find real effects. Another factor is the number of subjects—the fewer the subjects, the larger the difference necessary to find effects. A final factor is the willingness of the researcher to decide that caffeine made a difference when in fact it did not. The more the researcher is unwilling to make this kind of error, the larger the size of the difference necessary to decide that the effects are real. Hypothesis testing procedures incorporate these factors using a theoretical basis to help the researcher make a rational decision about whether the effects of caffeine are real or could be explained by chance.

At the conclusion of the experiment, then, one of two decisions will be made depending upon the size of the differences between the caffeine and no caffeine groups. The decision will either be that caffeine has an effect, making people more alert, or that chance factors (the composition of the group) could explain the result. The purpose of the hypothesis test is to eliminate false scientific conclusions as much as possible.

16-2 Rational Decisions

Most decisions require that an individual select a single alternative from a number of possible alternatives. The decision is made without knowing whether or not it is correct; that is, it is based on incomplete information. For example, a person either takes or does not take an umbrella to school based upon both the weather report and observation of outside conditions. If it is not currently raining, this decision must be made with incomplete information.

The concept of a decision by a rational man or woman is characterized by the use of a procedure that ensures that both the likelihood and the potential costs and benefits of all events are incorporated into the decision-making process. The procedure must be stated in such a fashion that another individual, using the same information, would make the same decision.

One is reminded of a "Star Trek" episode in which Captain Kirk is stranded on a planet without his communicator and is unable to get back to the *Enterprise*. Spock has assumed command and is being attacked by Klingons (who else?). Spock asks for and receives information about the location of the enemy, but is unable to act because he does not have complete information. Captain Kirk arrives at the last moment and saves the day because he can act on incomplete information.

This story goes against the concept of rational man. Spock, being a rational man, would not be immobilized by indecision. Instead, he would have selected the alternative that realized the greatest expected benefit given the information available. If complete information were required to make decisions, few decisions would be made by rational men and women. This is obviously not the case. The scriptwriter misunderstood Spock and rational man.

16-2a Effects

When a change in one thing is associated with a change in another, we have an **effect.** The changes may be either quantitative or qualitative, with the hypothesis testing procedure selected based upon the type of change observed. For example, if changes in sugar intake in a diet are associated with activity level in children, we say an effect occurred. In another case, if the distribution of political party preference (Republicans, Democrats, or Independents) differs for gender (male or female), then an effect is present. Much of the behavioral science is directed toward discovering and understanding effects.

The effects discussed in the remainder of this text are measured using various statistics including differences between means, a chi-square statistic computed from a contingency tables, and correlation coefficients.

16-3 General Principles

All hypothesis tests conform to similar principles and proceed with the same sequence of events. In almost all cases, the researcher wants to find statistically significant results. Failing to find statistically significant results means that the research will probably never be published, because few journals publish results that could be due to haphazard or chance findings. If research is not published, it is generally not very useful.

In order to decide that there are real effects, a model of the world is created in which there are no effects and the experiment is repeated an infinite number of times. The repetition is not "real," but rather is a "thought experiment" or mathematical deduction. The **sampling distribution** is used to create the model of the world with no effects and the study is repeated an infinite number of time.

The results of the single real experiment or study are compared to the theoretical model of no effects. If, given the model, the results are unlikely, then the model and the hypothesis of no effects generating the model are rejected and the effects are accepted as real. If the results could be explained by the model, the model must be retained, and no decision can be made about whether the effects were real.

Hypothesis testing is equivalent to the geometrical concept of hypothesis negation. That is, if one wants to prove that A (the hypothesis) is true, one first assumes that it isn't true. If it is shown that this assumption is logically impossible, then the original hypothesis is proven. In the case of hypothesis testing, the hypothesis may never be proven; rather, it is decided that the model of no effects is unlikely enough that the opposite hypothesis, that of real effects, must be true.

An analogous situation exists with respect to hypothesis testing in statistics. In hypothesis testing you want to show real effects of an experiment. By showing that the experimental results were unlikely, given that there were no effects, you may *decide* that the effects are, in fact, real. The hypothesis that there were no effects is called the **null hypothesis.** The symbol H_0 is used to abbreviate the null hypothesis in statistics. Note that, unlike geometry, we cannot *prove* the effects are real, rather we may *decide* the effects are real.

For example, suppose the probability model (distribution) in the following figure described the state of the world when there were no effects. In the case of Event A, the decision would be that the model could explain the results and the null hypothesis may true because Event A is fairly likely given that the model is true. On the other hand, if Event B occurred, the model would be rejected because Event B is unlikely, given the model.

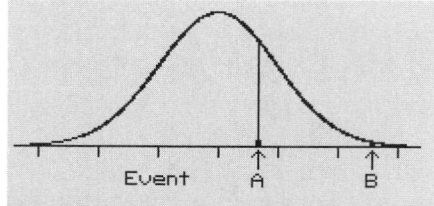

The probability of Event A is much higher than the probability of Event B.

16-3a The Model

The sampling distribution is a theoretical distribution of a sample statistic. It is used as a model of what would happen if

1. The null hypothesis were true (there really were no effects), and

2. The experiment were repeated an infinite number of times.

Because of its importance in hypothesis testing, the sampling distribution is the subject of the next chapter.

16-3b Probability

Probability theory essentially defines probabilities of simple events in algebraic terms and then presents rules for combining the probabilities of simple events into probabilities of complex events, given that certain conditions are present (assumptions are met). As such, probability theory is a mathematical model of uncertainty. It can never be "true" in an absolute sense, but may be more or less useful, depending upon how closely it mirrors reality.

Probabilities in an abstract sense are relative frequencies based on infinite repetitions. The probability of heads when a coin is flipped is the number of heads divided by the number of tosses as the number of tosses approaches infinity. In a similar vein, the probability of rain tonight is the proportion of times it rains given that conditions are identical to the conditions right now and they happen an infinite number of times. In neither the case of the coin nor the weather is it possible to "know" the exact probability of the event. Because of this, Henry Kyburg and Howard Smokler (*Studies in Subjective Probability*, 1964), among others, have argued that all probabilities are subjective and reflect a "degree of belief" about a relative frequency rather than a relative frequency.

Flipping a coin a large number of times is more intuitive than the exact weather conditions repeating themselves over and over again. Maybe that is why most texts begin by discussing coin tosses and drawing cards in an idealized game. The essential fact remains that it is impossible to flip a coin an infinite number of times. The true probability of obtaining heads or tails must always remain unknown. In a similar vein, it is impossible to manufacture a die that will have an exact probability of 1/6 for each side, although if enough care is taken the long-term results may be "close enough" that the casino will make money. The difficulty of computing a truly random sequence of numbers to use in simulations of probability experiments is well established (Ivars Peterson, *The Jungles of Randomness: A Mathematical Safari*, 1998).

The conceptualization of probabilities as unlimited relative frequencies has certain implications for probabilities of events that fall on the extreme ends of the continuum, however. The relative frequency of an impossible event must always remain at zero, no matter how many times it is repeated. The probability of getting an "arm" when flipping a coin must be zero, because although heads, tails, or an edge are possibilities, a coin has no "arm." An "arm" will never appear no matter how many times I flip a coin, thus its probability is zero.

In a like manner the probability of a certain event is one. The probability of a compound event such as obtaining heads, tails, or an edge when flipping a coin is a certainty, as one of these three outcomes must occur. No matter how many times a coin is flipped, one of the outcomes of this compound event must occur each time. Because any number divided by itself is one, the probability of a certain event is one.

The two extremes of zero and one provide the upper and lower limits to the values of probabilities. All values between these extremes can never be known exactly.

In addition to defining the nature of probabilities, probability theory also describes rules about how probabilities can be combined to produce probabilities of compound and conditional events. A **compound event** is a combination of simple events joined with either "and" or "or." For example, consider the statements, "We will win the conference football championship next season if the quarterback remains healthy and all the linemen pass this semester" and "All of the linemen will pass this semester if they study very hard or be very lucky." They both describe conditional events (we will win *if* x and y happen; the linemen will pass *if* x or y happens) that are dependent on compound events (the quarterback must stay healthy AND the linemen must pass this semester; the linemen must study hard OR be lucky). In other words, the compound events must be true before the condition can be true. A conditional event also could employ the term "given." For example, "The university football team will win the conference football championship next season given the quarterback remains healthy and all the linemen pass this semester." The condition following the word "given" must be true before the condition before the "given" takes effect.

Combining Probabilities of Independent Events

The probability of a compound event described by the word "and" is the product of the simple events if the simple events are independent. To be independent, two events cannot possibly influence each other. For example, as long as you are willing to assume that the events of the quarterback remaining healthy and the linemen all passing are independent, then the probability of winning the conference football championship can be calculated by multiplying the probabilities of each of the separate events together. For example, if the probability of the quarterback remaining healthy is .6 and the probability of all the linemen passing this semester is .2, then the probability of winning the conference championship is .6 * .2 or .12. This relationship can be written in symbols as follows:

P(A and B) = P(A) * P(B) if A and B are independent events

Combining Probabilities Using "Or"

If the compound event can be described by two or more events joined by the word "or," then the probability of the compound event is the sum of the probabilities of the individual events minus the probability of the joint event. For example, the probability of all the linemen passing would be the sum of the probability of all studying very hard plus the probability of all being very lucky, minus the probability of all studying very hard and all being very lucky. For example, suppose that the probability of all studying very hard was .15, the probability of all being very lucky was .0588, and the probability of all studying very hard and all being very lucky was .0088. The probability of all passing would be .15 + .0588 − .0088 = .20. In general, the relationship can be written as follows:

P(A or B) = P(A) + P(B) − P(A and B)

Conditional Probabilities

A **conditional probability** is the probability of an event given another event is true. The probability that the quarterback will remain healthy given that he stretches properly at practice and before game time would be a conditional probability. By definition a conditional probability is the probability of the joint event divided by the probability of the conditional event. The probability that the quarterback will remain healthy given that he stretches properly at practice and before game time would be the probability of the quarterback both remaining healthy and stretching properly divided by the probability of stretching properly. Suppose the probability of stretching properly is .8 and the probability of both stretching properly and remaining healthy is .55. The conditional probability of remaining healthy given that he stretched properly would be .55/8 = .6875. The "given" is written in probability theory as a vertical line (|), such that the preceding could be written as

P(A|B) = P(A and B) / P(B)

Conditional probabilities can be combined into a very useful formula called Bayes's Rule. This equation describes how to modify a probability given information in the form of conditional probabilities:

P(A|B) = (P(B|A) * P(A)) / (P(B|A) * P(A) + P(B|not A) * P(not A))

where A and B are any events whose probabilities are not 0 and 1.

Suppose that an instructor randomly picks a student from a class where males outnumber females two to one. What is the probability that the selected student is a female? Given the ratio of males to females, this probability could be set to 1/3 or .333. This probability is called the prior probability and would be represented in the previous equation as P(A). In a similar manner, the probability of the student being a male, P(not A), would be 2/3 or .667. Suppose additional information was provided about the selected student, that the shoe size of the person selected was 7.5. Often it is possible to compute the conditional probability of B given A or in this case, the probability of a size 7.5 given the person was a female. In a like manner, the probability of B given not A can often be calculated; in this case the probability of a size 7.5 given the person was a male. Suppose the former probability is .8 and the latter is .1. The likelihood of the person being a female given a shoe size of 7.5 can be calculated using Bayes's Rule as follows:

$$P(A|B) = (P(B|A) * P(A)) / (P(B|A) * P(A) + P(B|not A) * P(not A))$$

$$= (.8 * .333) / (.8 * .333 + .1 * .667)$$

$$= .2664 / .3331 = .7998$$

The value of P(A|B) is called a posterior probability and in this case the probability of the student being a female given a shoe size of 7.5 is fairly high at .7998. The ability to recompute probabilities based on data is the foundation of a branch of statistics called **Bayesian Statistics.**

This set of rules barely scratches the surface when considering the possibilities of probability models.

16-3c Including Cost in Making Decisions with Probabilities

Including cost as a factor in the equation can extend the usefulness of probabilities as an aid in decision-making. This is the case in a branch of statistics called utility theory, which includes a concept called **utility** in the equation. Utility is the gain or loss experienced by a player depending upon the outcome of the game and can be symbolized with a "U." Usually utility is expressed in monetary units, although there is no requirement that it must be. The symbol U(A) would be the utility of outcome A to the player of the game. A concept called expected utility would be the result of playing the game an infinite number of times. In its simplest form, expected utility is a sum of the products of probabilities and utilities:

Expected Utility = P(A) * U(A) + P(not A) * U(not A)

Suppose someone was offered a chance to play a game with two dice. If the dice totaled 6 or 8, the player would receive $70, otherwise he or she would pay $30. The utility to the player is plus $70 for A and minus $30 for not A. The probability of rolling a 6 or 8 is 10/36 = .2778, while the probability of some other total is .7222. Should the player consider the game? Using expected utility analysis, the expected utility would be:

Expected Utility = (.2778 * 70) + (.7222 * (-30)) = -2.22

Since the expected utility is less than 0, indicating a loss over the long run, **expected utility theory** would argue against playing the game. Again this illustration just barely scratches the surface of a very complex and interesting area of study and the reader is directed to other sources for further study. In particular, the area of game theory holds a great deal of promise.

You should be aware that the preceding analysis of whether or not to play a given game based on expected utility assumes that the dice are "fair," that is, each face is equally likely to turn up. To the extent the fairness assumption is incorrect—for example using weighted dice—the theoretical analysis will also be incorrect. Going back to the original definition of probabilities, that of a relative frequency given an infinite number of possibilities, it is never possible to know the exact probability of any event.

16-3d Using Probability Models in Science

Does this mean that all the preceding is useless? Absolutely not! It does mean, however, that probability theory and probability models must be viewed within the larger framework of model-building in science. The "laws" of probability are a formal language model of the world that, like algebra and numbers, exists as symbols and relationships between symbols. They have no meaning in and of themselves and belong in the portion of the model-building paradigm that is circled in the following figure:

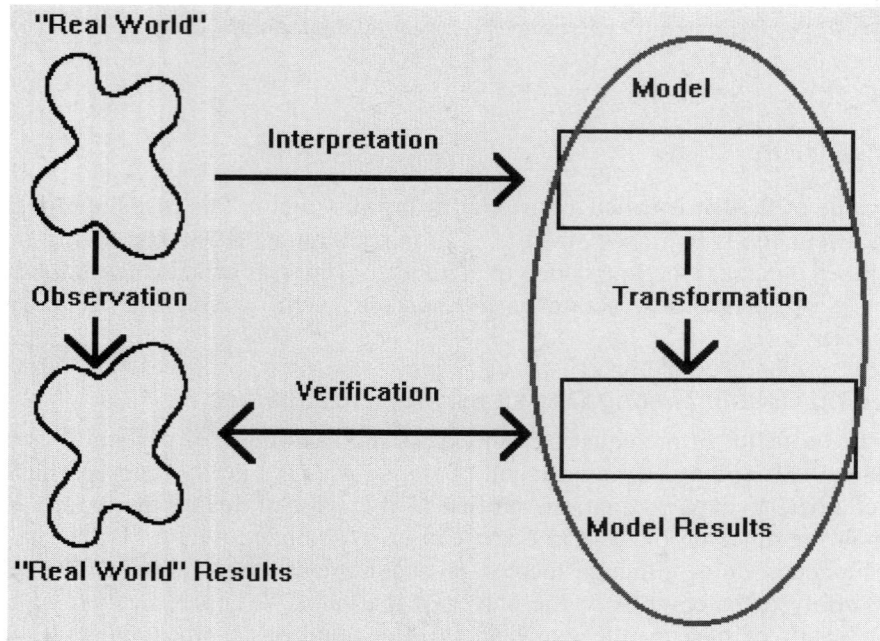

The model-building process—transforming the model.

As with numbers and algebraic operators, the symbols within the language must be given meaning before the models become useful. In this case, interpretation implies that numbers are assigned to probabilities based on rules. The circled part of the following figure illustrates the portion of the model-building process that now becomes critical:

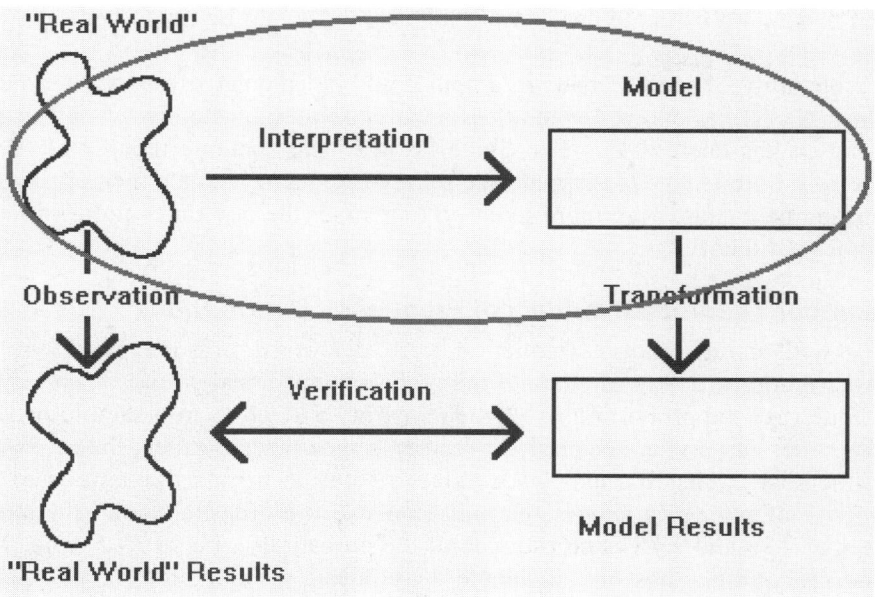

The model-building process—creating the model.

16-4 Establishing Probabilities

There are a number of different ways to estimate probabilities. Each has advantages and disadvantages, and some have proven more useful than others. Just because a number can be assigned to a given probability symbol, however, does not mean that the number is the "true" probability.

16-4a Equal Likelihood

When there is no reason to believe that any outcome is more or less likely than any other outcome, then the solution is to assign all outcomes an equal probability. For example, since there is no reason to believe that heads is more likely than tails, a value of .5 is assigned to each when a coin is flipped. In a similar manner, if there is no reason to believe that one card is more likely to be picked than any other, then a probability of 1/52 or .0192 is assigned to every card in a standard deck.

This system does not work when there is reason to believe that one outcome is more likely than another. For example, setting a probability of .5 that it will be snowing outside in an hour is not reasonable. There are two alternatives—it will either be snowing or it won't—but equal probabilities are not tenable because it is sunny and 60 degrees outside my office right now and I have reason to believe that it will not be snowing in an hour.

16-4b Relative Frequency

The relative frequency of an event in the past can be used as an estimate of its probability. For example, the probability of a student succeeding in a given graduate program could be calculated by dividing the number of students having actually finished the program by the number of students admitted in the past. Establishing probabilities in this fashion assumes that conditions in the past will continue into the future, generally a fairly safe bet. The greater the number of observations, the more stable the estimate based on relative frequency. For example, the probability of heads for a given coin toss could be calculated by dividing the number of heads by the number of tosses. An estimate based on 10,000 tosses would be much better than one based on 10 tosses.

The probability of snow outside in a hour could be calculated by dividing the number of times in the past that it has snowed when the temperature an hour before was 60 degrees by the number of times it has been 60 degrees. Since I don't have accurate records of such events, I would have to rely on memory to estimate the relative frequency. Since memory seems to work better for outstanding events, I am more likely to remember the few times it did snow in contrast to the many times it did not.

16-4c Area under Theoretical Models of Frequency Distributions

The problems with using relative frequency were discussed in some detail in Chapter 5, "Frequency Distributions." If an estimate of the probability of females who wear size 7.5 shoes is needed, one could use the proportion of women wearing a size 7.5 in a sample of women. The problem is that unless a very large sample of women's shoe sizes is taken, the relative frequency of any one shoe size is unstable and inaccurate. A solution to this dilemma is to construct a theoretical model of women's shoe sizes and then use the area under the theoretical model between values of 7.25 and 7.75 as an estimate of the probability of a size 7.5 shoe. This method of establishing probabilities has the advantage of requiring a much smaller sample to estimate relatively stable probabilities. Its disadvantage is that the probability estimation is several steps removed from the relative frequency, requiring both the selection of the model and the estimation of the parameters of the model. Fortunately, selecting the correct model and estimating parameters of the models is a well-understood and thoroughly studied topic in statistics.

Area under theoretical models of distributions is the method that classical hypothesis testing employs to estimate probabilities. A major part of an intermediate course in mathematical statistics is the theoretical justification of the models that are used in hypothesis testing.

16-4d Subjective Probabilities

A controversial method of estimating probabilities is to simply ask people to state their degree of belief as a number between zero and one and then treat that number as a probability. A slightly more sophisticated method is to ask the odds the person would be willing to take in order to place a bet. Probabilities obtained using these and other subjective methods are called **subjective probabilities.** If someone was told, "Give me a number between zero and one, where zero is impossible and one is certain, to describe the likelihood of Jane Student finishing the graduate program," that number would be a subjective probability.

Subjective probabilities have the greatest advantage in that they are intuitive and easy to obtain. People use subjective probabilities all the time to make decisions. For example, my decision about what to wear when I leave the house in the morning is partially based on what I think the weather will be like an hour from now. A decision on whether or not to take an umbrella is based partly on the subjective probability of rain. A decision to invest in a particular company in the stock market is partly based on the subjective probability that the company will increase in value in the future.

The greatest disadvantage of subjective probabilities is that people are notoriously bad at estimating the likelihood of events, especially rare or unlikely events. Memory is selective. Human memory is poorly structured to answer queries such as estimating the relative frequency of snow an hour after the temperature was 60 degrees Fahrenheit and likely to be influenced by significant, but rare, events. If asked to give a subjective probability of snow in an hour, the resulting probability estimate would be a **compound probability** resulting from a large number of conditional probabilities, such as the latest weather report, the time of year, the current temperature, and intuitive feelings.

16-5 Inaccurate Estimates of Probabilities and Their Effect

Subjective probability estimates are influenced by emotion. In assessing the likelihood of your favorite baseball team winning the pennant, feelings are likely to intervene and make the estimate larger that reality would suggest. Bookmakers (bookies) everywhere bank on such human behavior. In a similar manner, people are likely to assess the likelihood of experimental methods to cure currently incurable diseases as much higher than they actually are, especially when they have an incurable disease. The foundation of lotteries is an overestimate of the probability of winning. Almost every winner in a casino is celebrated by lights flashing and bells ringing, causing patrons to maintain a general overestimate of the probability of winning.

People have a difficult time assessing risk and responding appropriately, especially when the probabilities of the events are low. In the late 1980s people were canceling overseas travel because of threats of terrorist attacks. John Allen Paulos (*Innumeracy: Mathematical Illiteracy and its Consequences*, 1988) estimates that the likelihood of being killed by terrorists in any given year is one in 1,600,000 while the chance of dying in a car crash in the same time frame is one in only 5,300. Yet people still refuse to use seat belts.

When people are asked to estimate the probability of some event, the event occurs, and then the same people are asked what their original probabilities were, they almost inevitably inflate them in the direction of the event. For example, suppose people were asked to give the probability that a particular candidate would win an election, the candidate won, and then the same people were asked to repeat the probability that they originally presented. In almost all cases, the probability would be higher than the original probability. This well-established phenomenon is called hindsight bias (Anders Wenman, Peter Juslin, and Mats Bjorkman in a 1998 *Journal of Experimental Psychology* article).

Since most subjective probability estimates are compound probabilities, humans also have a difficult time combining simple probabilities into compound probabilities. Some of the difficulty has to do with a lack of understanding about independence and mutual exclusivity necessary to multiply and add probabilities. If a couple has three children, all boys, the probability of the next child being a boy is approximately .5, even though the probability of having four boys is $.5^4$ or .0625. The correct probability is a conditional probability of having a boy given that they already had three boys.

Another difficulty with probabilities has to do with a misunderstanding about conditional probability. When subjects were asked to rank potential occupations of a person described by a former neighbor as "very shy and withdrawn, invariably helpful, but with little interest in people, or in the world of reality. A meek and tidy soul, he has a need for order and structure, and a passion for detail," Amos Tversky and Daniel Kahneman (in a 1974 *Science* article) found that they inevitably categorize him as a librarian rather than a farmer. People fail to take into account that the base rate or prior probability of being a farmer is much higher than that of being a librarian.

Using this and other illustrations of systematic and predictable errors made by humans is assessing probabilities, Tversky and Kahneman argue that reliance on subjective probabilities to assign values to symbols used within probability theory will inevitably lead to logical contradictions.

16-6 Using Probabilities

16-6a In the Long Run

The casinos in Las Vegas and around the world are testaments that probability models work as advertised. Insurance companies seldom go broke. There is no question that probability models work if care is used in their construction and the user has the ability to participate for the long run. These models are so useful that Peter Bernstein (*Against the Gods*, 1996) has claimed, "The revolutionary idea that defines the boundary between modern times and the past is the mastery of risk: the notion that the future is more than a whim of the gods and that men and women are not passive before nature."

In hypothesis testing, probability models are used to control the proportion of times a researcher claims to have found effects when, in fact, the results were due to chance or haphazard circumstances. Because the science as a whole is able to participate in the long run, these models have been successfully applied with the result that only a small proportion of published research is the result of chance, coincidence, or haphazard events.

16-6b In the Short Run

Most of the decisions that are made in real life are made without the ability to view the results in the long run. An undergraduate student decides to apply to a given graduate school based upon an assessment of the probability of a favorable outcome and the benefits of attending that particular school. There is generally no opportunity to apply to the same program over and over again and observe the results. Probability models have limited value in these situations because of the difficulties in estimating probabilities with any kind of accuracy.

Personally, I use expected utility theory in justifying not playing the lottery or gambling in casinos. If the expected value is less than zero, I don't play. That doesn't explain why I carry insurance on my house, other than that the bank requires it for a mortgage, and my health, other than that the university provides it as part of my benefits.

It has been fairly well established that probability and utility theory are not accurate normative models of how people actually make decisions. In a 1998 article in *Journal of Economic Issues*, John Harvey argues that people use a variety of heuristics, or rules of thumb, to make decisions about the world. An awareness and use of probability and utility theory have the potential benefit of making the people much better decision-makers and are worthy of further study.

Summary

This chapter discussed the basics of hypotheses, hypothesis testing, and probability theory. You learned that hypothesis tests are procedures for making rational decisions about the reality of effects, and that an effect is a change in one thing associated with a change in another. In order to perform a hypothesis test, you first create a model of what the world would look like if there were no effects and then repeat the study an infinite number of times. You compare the results of your single study with the model. If the results of the single study are unlikely enough given the model, you reject the model and the hypothesis of no effects that generated the model and accept the hypothesis that the effects are real.

Sampling distributions as theoretical distributions of sample statistics were used to create the model for hypothesis testing.

You learned that probability theory defines probabilities of simple events in algebraic terms and then presents rules for combining the probabilities of simple events into probabilities of complex events given that certain conditions are present (assumptions are met). As such, probability theory is a mathematical model of uncertainty.

Conditional probabilities, including cost as a factor in the probability equation (utility theory), and the need to view probability theory and probability models in the larger framework of model building also were discussed. And you learned three basic ways to establish probabilities: equal likelihood, relative frequency of an event in the past, and subjective probabilities.

Some of the topics were only introduced in this chapter and are covered in more detail in other chapters in the text.

Chapter

17

The Sampling Distribution

Key Terms

Central Limit Theorem	population distribution	sample statistics
estimator	probability models	standard error
parameters	sample distribution	unbiased estimator

The sampling distribution is a theoretical distribution of a sample statistic. While the concept of a distribution of a set of numbers is intuitive for most students, the concept of a distribution of a set of statistics is not. Therefore, let's review distributions before we discuss the sampling distribution.

17-1 The Sampling Distribution

The **sample distribution** is the distribution resulting from the collection of actual data. A major characteristic of a sample is that it contains a finite (countable) number of scores, the number of scores represented by the letter N. For example, suppose that the following data were collected:

Sample Data																				
32	35	42	33	36	38	37	33	38	36	35	34	37	40	38	36	35	31	37	36	33
36	39	40	33	30	35	37	39	32	39	37	35	36	39	33	31	40	37	34	34	37

These numbers constitute a sample distribution. Using the procedures discussed in Chapter 5, "Frequency Distributions," the following histogram can be constructed to picture this data:

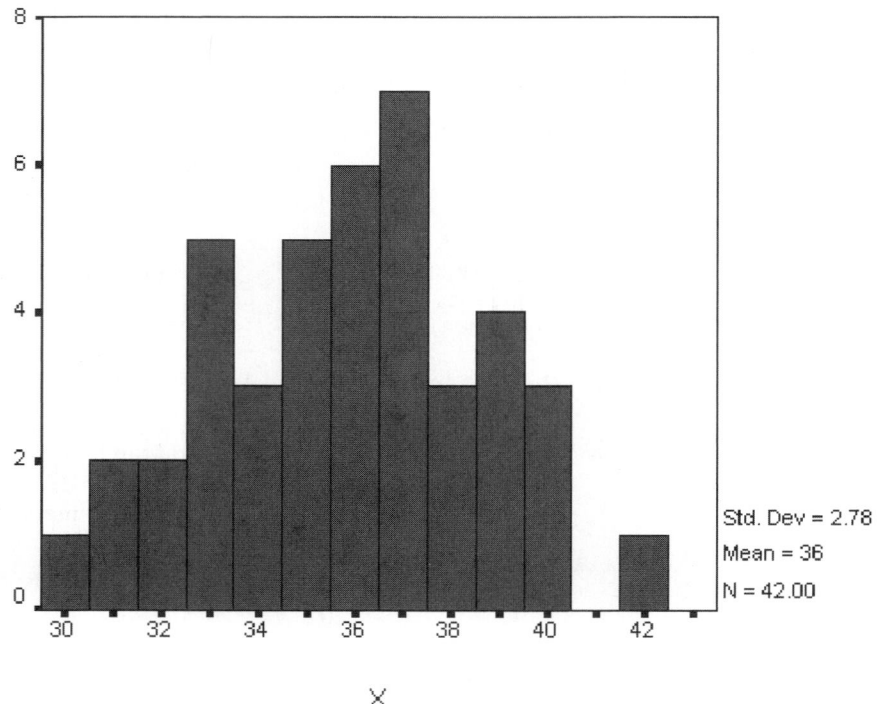

Std. Dev = 2.78
Mean = 36
N = 42.00

Visualizing the sample distribution with a histogram.

In addition to the frequency distribution, the sample distribution can be described with numbers, called statistics. Examples of statistics are the mean, median, mode, standard deviation, range, and correlation coefficient, among others. Statistics, and procedures for computing statistics, were discussed in detail in earlier chapters.

If a different sample were taken, different scores would result. The relative frequency polygon would be different, as would the statistics computed from the second sample. However, there would also be some consistency in that while the statistics would not be exactly the same, they would be similar. To achieve order in this chaos, statisticians have developed probability models.

17-2 Probability Models: Population Distributions

Probability models exist in a theoretical world where complete information is available. As such, they can never be known except in the mind of the mathematical statistician. If an infinite number of infinitely precise scores were taken, the resulting distribution would be a probability model of the population.

The probability model may be described with pictures (graphs), which are analogous to the relative frequency polygon of the sample distribution. The two graphs below illustrate two types of probability models, the uniform distribution and the normal curve.

A uniform distribution.

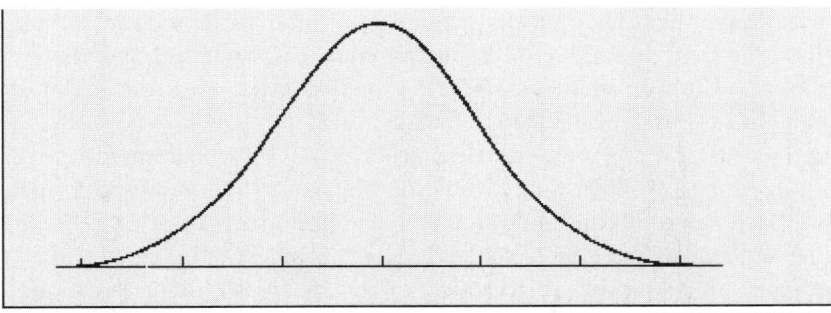

A normal distribution.

As discussed in Chapter 9, "The Normal Curve," probability distributions are described by mathematical equations that contain parameters. **Parameters** are variables that change the shape of the probability model. By setting these parameters equal to numbers, a member of that probability model family of models results.

A critical aspect of statistics is the estimation of parameters with sample statistics. **Sample statistics** are used as **estimators** of the corresponding parameters in the population model. For example, the mean and standard deviation of the sample are used as estimates of the corresponding population parameters μ and σ. Mathematical statistics texts devote considerable effort to defining what is a good or poor parameter estimation procedure.

17-3 The Sampling Distribution

Note the "ing" on the end of *sample* in sampling distribution. It looks and sounds similar to the sample distribution but, in reality, the concept is much closer to a population model.

The sampling distribution is a theoretical distribution of a sample statistic. It is a model of a distribution of scores, like the **population distribution,** except that the scores are not raw scores, but statistics. It is a thought experiment: "What would the world be like if a person repeatedly took samples of size N from the population distribution and computed a particular statistic each time?" The resulting distribution of statistics is called the sampling distribution of that statistic.

For example, suppose that a sample of size sixteen (N = 16) is taken from some population. The mean of the sixteen numbers is computed. Next, a new sample of sixteen is taken, and the mean is again computed. If this process were repeated an infinite number of times, the distribution of the now infinite number of sample means would be called the sampling distribution of the mean.

Every statistic has a sampling distribution. For example, suppose that instead of the mean, medians were computed for each sample. The infinite number of medians would be called the sampling distribution of the median.

Just as the population models can be described with parameters, so can the sampling distribution. The expected value (analogous to the mean) of a sampling distribution will be represented here by the symbol μ (mu). The μ symbol is often written with a subscript to indicate which sampling distribution is being discussed. For example, the expected value of the sampling distribution of the mean is represented by the symbol $\mu_{\bar{X}}$, that of the median by μ_{M_d}, and so on. The value of $\mu_{\bar{X}}$ can be thought of as the mean of the distribution of means. In a similar manner the value of μ_{M_d} is the mean of a distribution of medians. They are not really means, because it is not possible to find a mean when N = ∞, but they are the mathematical equivalent of a mean.

Using advanced mathematics, in a thought experiment, the theoretical statistician often discovers a relationship between the expected value of a statistic and the model parameters. For example, it can be proven that the expected value of both the mean and the median, $\bar{X}$ and M_d, is equal to μ_X. When the expected value of a statistic equals a population parameter, the statistic is called an **unbiased estimator** of that parameter. In this case, both the mean and the median would be an unbiased estimator of the parameter μ_X.

A sampling distribution may also be described with a parameter corresponding to a variance, symbolized by σ^2. The square root of this parameter is given a special name, the **standard error.** Each sampling distribution has a standard error. In order to keep them straight, each has a name tagged on the end of "standard error" and a subscript on the σ symbol. The standard deviation of the sampling distribution of the mean is called the standard error of the mean and is symbolized by $\sigma_{\bar{X}}$. Similarly, the standard deviation of the sampling distribution of the median is called the standard error of the median and is symbolized by σ_{M_d}.

In each case the standard error of statistics describes the degree to which the computed statistics will differ from one another when calculated from a sample of similar size and selected from similar population models. The larger the standard error, the greater the difference between the computed statistics. Consistency is a valuable property to have in the estimation of a population parameter, as the statistic with the smallest standard error is preferred as the estimator of the corresponding population parameter, everything else being equal. Statisticians have proven that, in most cases, the standard error of the mean is smaller than the standard error of the median. Because of this property, the mean is the preferred estimator of μ_X.

17-4 The Sampling Distribution of the Mean

The sampling distribution of the mean is a distribution of sample means. This distribution may be described with the parameters $\mu_{\bar{X}}$ and $\sigma_{\bar{X}}$.

These parameters are closely related to the parameters of the population distribution, with the relationship being described by the **Central Limit Theorem.** The Central Limit Theorem essentially states that the mean of the sampling distribution of the mean ($\mu_{\bar{X}}$) equals the mean of the population model (μ_X) and that the standard error of the mean ($\sigma_{\bar{X}}$) equals the standard deviation of the population model (σ_X) divided by the square root of N as the sample size gets infinitely larger (N—>∞). In addition, the sampling distribution of the mean will approach a normal distribution. The following equations summarize these relationships:

$$\mu_{\bar{X}} = \mu_X$$
$$\sigma_{\bar{X}} = \frac{\sigma_X}{\sqrt{N}}$$

Critical relationships between the parameters of the sampling distribution of the mean and the model.

The astute student has probably noticed, however, that the sample size would have to be very large (∞) in order for these relationships to hold true. In theory, this is fact; in practice, an infinite sample size is impossible. The Central Limit Theorem is very powerful. In most situations encountered by behavioral scientists, this theorem works reasonably well with an N greater than 10 or 20. Thus, it is possible to closely approximate what the distribution of sample means looks like, even with relatively small sample sizes.

The importance of the Central Limit Theorem to statistical thinking cannot be overstated. Most of hypothesis testing and sampling theory is based on this theorem. In addition, it provides a justification for using the normal curve as a model for many naturally occurring phenomena. If a trait, such as intelligence, can be thought of as a combination of relatively independent events, in this case both genetic and environmental, then it would be expected that the trait would be normally distributed in a population.

Summary

The sampling distribution, a theoretical distribution of a sample statistic, is a critical component of hypothesis testing. The sampling distribution allows the statistician to hypothesize about what the world would look like if a statistic was calculated an infinite number of times.

A sampling distribution exists for every statistic that can be computed. Like models of relative frequency distributions, sampling distributions are characterized by parameters, two of which are the theoretical mean and standard deviation. The theoretical standard deviation of the sampling distribution is called the standard error and describes how much variation can be can be expected given certain conditions are met, such as a particular sample size.

Of considerable importance to statistical thinking is the sampling distribution of the mean, a theoretical distribution of sample means. A mathematical theorem, called the Central Limit Theorem, describes the relationship of the parameters of the sampling distribution of the mean to the parameters of the probability model and sample size. The Central Limit Theorem is the theoretical foundation of the gaming industry and insurance companies. The Central Limit Theorem is also the foundation for hypothesis tests that measure the size of the effect using means.

18

Testing Hypotheses about Single Means

Key Terms

alpha   exact significance level

18-1 The Head-Start Study

Suppose an educator had a theory which argued that a great deal of learning occurs before children enter grade school or kindergarten. This theory explained that socially disadvantaged children start school intellectually behind other children and are never able to catch up. In order to remedy this situation, he proposes a head-start program, which starts children in a school situation at ages three and four.

A politician reads this theory and feels that it might be true. However, before she is willing to invest the billions of dollars necessary to begin and maintain a head-start program, she demands that the educator demonstrate that the program really does work. At this point, the educator calls for the services of a researcher and statistician.

Because this is a fantasy, the following research design would never be used in practice, but it will illustrate the procedure and the logic underlying the hypothesis test. A more appropriate design will be discussed later in the text.

A random sample of 64 children is taken from the population of all four-year-old children. The children in the sample are all enrolled in the head-start program for a year, at the end of which time they are given a standardized intelligence test. The mean I.Q. of the sample is found to be 103.27.

On the basis of this information, the educator wants to begin a nationwide head-start program. He argues that the average I.Q. in the population is 100 ($\mu = 100$) and that 103.27 is greater than that. Therefore, the head-start program had an effect of about $103.27 - 100$ or 3.27 I.Q. points. As a result, the billions of dollars necessary for the program would be well invested.

The statistician, being in this case the devil's advocate, is not ready to act so hastily. She wants to know whether chance could have caused the large mean. In other words, the head start program doesn't make a bit of difference. The mean of 103.27 was obtained because the sixty-four students selected for the sample were slightly brighter than average. She argues that this possibility must be ruled out before any action is taken. If it is not possible to completely rule out this possibility, she argues that although possible, the likelihood must be small enough that the risk of making a wrong decision outweighs possible benefits of making a correct decision.

To determine if chance could have caused the difference, the hypothesis test proceeds as a thought experiment. First, the statistician assumes that there were no effects; in this case, that the head-start program didn't work. She then creates a model of what the world would look like if the experiment were performed an infinite number of times under the assumption of no effects. The sampling distribution of the mean is used as this model. The reasoning goes something like this:

Population model assuming no effects:

$$\mu_X = 100$$
$$\sigma_X = 16$$

Sampling distribution assuming no effects, N = 64:

$$\mu_{\bar{X}} = \mu_X = 100$$
$$\sigma_{\bar{X}} = \frac{\sigma_X}{\sqrt{N}} = \frac{16}{\sqrt{64}} = \frac{16}{8} = 2.0$$

Results of the study:

$$N = 64$$

$$\bar{X} = 103.27$$

The researcher then compares the results of the actual experiment with those expected from the model, given there were no effects and the experiment was repeated an infinite number of times. The Probability Calculator is used to find the probability of the results of the study given the model of no effects. The probability of finding a result equal to or greater than the actual result is called an **exact significance level.** It is standard practice to find a two-tailed significance level, taking the total area under the curve both above the score and below a mirror image of the score. In the illustration that follows, the probability of a score greater than 103.27 or less than 96.73 is .102, or greater than one in 10. The researcher concludes that the model of no effects was not unlikely enough to reject and therefore could explain the results.

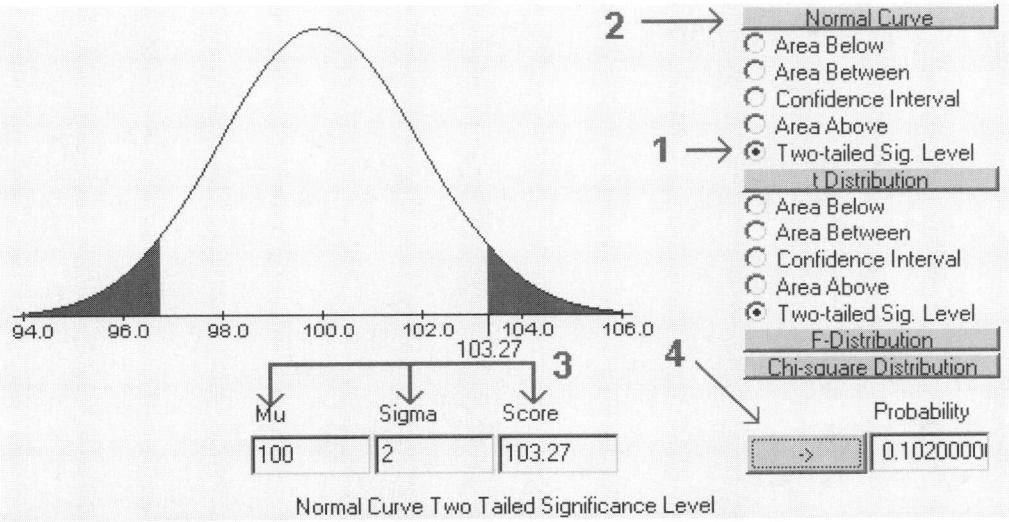

Normal Curve Two Tailed Significance Level

Results of the study compared to theoretical sampling distribution assuming there were no effects.

Of critical importance is the question, "How unlikely do the results have to be before they are unlikely enough?" The researcher must answer this question before the analysis is started, and the answer is stated as a probability, called **alpha.** Alpha is the probability of rejecting the null hypothesis given that the null hypothesis is true, or in other words, the probability of deciding that the effects are real when in fact chance caused the results. An almost universally accepted default value for alpha is .05, although a later chapter deals with the reasons for selections of alternative values for alpha. Unless stated otherwise, the value for alpha will be assumed to be .05.

If the exact significance level is greater than alpha, then the decision of the hypothesis test will always be to retain the null hypothesis. If the exact significance level is less than alpha, then the null hypothesis will be rejected and the alternative hypothesis accepted. In this case, because the value of the exact significance level (.102) is greater than alpha (.05), the null hypothesis must be retained. Therefore, because chance could explain the results, the educator was premature in deciding that head-start had a real effect.

18-2 The Head-Start Study Redone

Suppose that the researcher changed the experiment. Instead of a sample of sixty-four children, the sample was increased to N = 400 four-year-old children. Furthermore, this sample had the same mean ($\bar{x}$) at the conclusion as had the previous study. The statistician must now change the model to reflect the larger sample size.

Population model assuming no effects:

$$\mu_X = 100$$
$$\sigma_X = 16$$

Sampling distribution assuming no effects, N = 400:

$$\mu_{\bar{X}} = \mu_X = 100$$
$$\sigma_{\bar{X}} = \frac{\sigma_X}{\sqrt{N}} = \frac{16}{\sqrt{400}} = \frac{16}{20} = 0.8$$

Results of the study:

$$N = 400$$
$$\bar{X} = 103.27$$

The conclusion reached by the statistician states that it is highly unlikely the model could explain the results. The model of chance is rejected and the reality of effects accepted. Why? The mean that resulted from the study fell in the tail of the sampling distribution, as this figure shows:

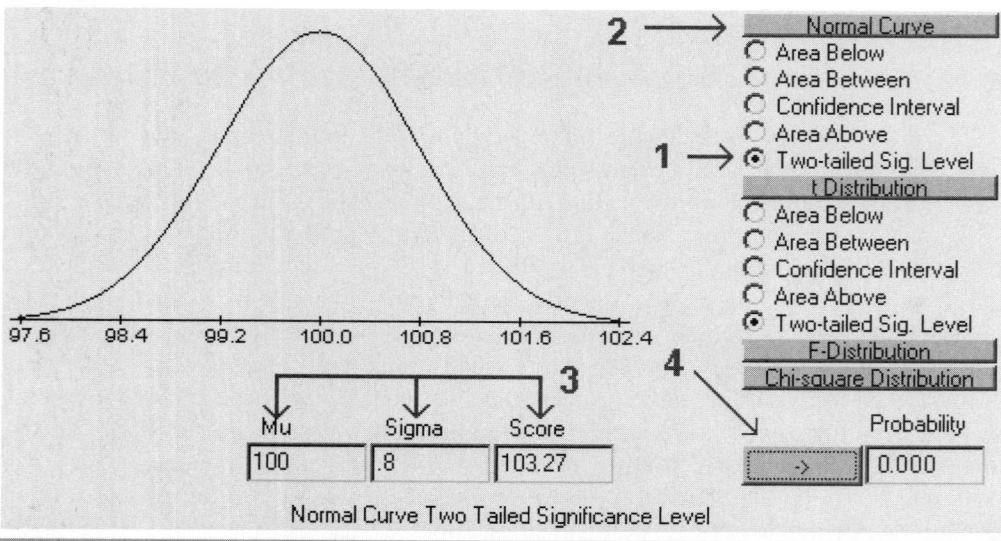

Normal Curve Two Tailed Significance Level

Results of the study compared to theoretical sampling distribution assuming there were no effects.

The exact significance level (0.0 in this case) is certainly less than alpha, so the null hypothesis must be rejected and the alternative hypothesis accepted. The exact significance level is never exactly equal to zero, as there is always some area under the tails of the curve no matter how far in the tail the score falls. Often the area is so small that it doesn't register in three decimal places and will be shown as .000. In these cases, unless alpha is set to an extremely small value, the null hypothesis will always be rejected.

The different conclusions reached in these two studies may seem contradictory. A little reflection, however, reveals that the second study was based on a much larger sample size (400 versus 64). As such, the researcher is rewarded for doing more careful work and taking a larger sample. The sampling distribution of the mean specifies the nature of the reward.

At this point it should also be pointed out that we are discussing statistical significance: whether or not the results might have occurred by chance. The second question, that of practical significance, occurs only after an affirmative decision about the reality of the effects. The practical significance question is tackled by the politician, who must decide whether the effects are large enough to be worth the money to begin and maintain the program. Even though head-start works, the money may be better spent in programs for health care of the aged or for more nuclear submarines. In short, this is a political and practical decision made by people and not statistical procedures.

Summary

This chapter illustrated a significance test comparing a single mean to a population parameter (μ_X).

A model of what the world looks like—given that there were no effects and that the experiment was repeated an infinite number of times—was created using the sampling distribution of the mean. The mean of the experiment was compared to the model to decide whether the effects were due to chance or whether another explanation was necessary (the effects were real). In the first case, the decision was made to retain the model. It could explain the results. In the second case, the decision was to reject the model and accept the reality of the effect.

Finally, the difference between statistical significance and practical significance was discussed.

Chapter

19

The T-Test

Key Terms

Before an experiment is performed, the question of experimental design must be addressed. **Experimental design** refers to the manner in which the experiment will be set up, specifically the way the treatments will be administered to subjects. Treatments will be defined as quantitatively or qualitatively different levels of experience. For example, in an experiment on the effects of caffeine, the treatment levels might be exposure to different amounts of caffeine, from 0 to .0375 milligram. In a very simple experiment there are two levels of treatment; none, called the **control condition,** and some, called the **experimental condition.**

19-1 Experimental Designs

The type of analysis or hypothesis test used is dependent upon the type of experimental design employed. The two basic types of experimental designs are *crossed* and *nested*.

In a **crossed design,** each subject sees each level of the **treatment conditions.** In a very simple experiment, such as one that studies the effects of caffeine on alertness, each subject would be exposed to both a caffeine condition and a no-caffeine condition. For example, using the members of a statistics class as subjects, the experiment might be conducted in the following manner:

> On the first day of the experiment, the class is divided in half, with one half of the class getting coffee with caffeine and the other half getting coffee without caffeine. A measure of alertness is taken for each individual, such as the number of yawns during the class period. On the second day the conditions are reversed; that is, the individuals who received coffee with caffeine on the first day are now given coffee without, and vice-versa. The size of the effect will be the difference of alertness on the days with and without caffeine.

The distinguishing feature of crossed designs is that each individual will have more than one score. The effect occurs within each subject, thus these designs are sometimes referred to as *within subjects* designs. In SPSS, the analysis is called a Paired-Samples T-Test.

Crossed designs have two advantages: they generally require fewer subjects (because each subject is used a number of times in the experiment), and they are more likely to result in a significant effect, given the effects are real.

Crossed designs also have some disadvantages. For one, the experimenter must be concerned about carryover effects. For example, individuals not used to caffeine may still feel the effects of caffeine on the second day, even though they did not receive the drug. Secondly, the first measurements taken may influence the second. For example, if the measurement of interest was a score on a statistics test, taking the test once may influence performance the second time the test is taken. Finally, the assumptions necessary when more than two treatment levels are employed in a crossed design may be restrictive.

In a **nested design,** each subject receives one, and only one, treatment condition. The critical difference in the simple experiment described earlier is that the experiment would be performed on a single day, with half the individuals receiving coffee with caffeine and half receiving coffee without caffeine. The size of effect in this case is determined by comparing average alertness between the two groups.

The major distinguishing feature of nested designs is that each subject has a single score. The effect, if any, occurs between groups of subjects and thus the name *between subjects* is sometimes given to these designs. In SPSS the analysis of nested designs with two groups is called Independent-Samples T-Test.

The relative advantages and disadvantages of nested designs are opposite those of crossed designs: carryover effects are not a problem, as individuals are measured only once, but the number of subjects needed to discover effects is greater than with crossed designs.

Some treatments are nested by their nature. The effect of gender, for example, is necessarily nested. One is either a male or a female, but not both. Current religious preference is another example. Effects that rely on a pre-existing condition are sometimes called demographic or blocking factors, and subjects will always be nested within these factors.

19-2 Crossed Designs

As discussed earlier, a crossed design occurs when each subject sees each **treatment** level, that is, when there is more than one score per subject. The purpose of the analysis is to determine if the effects of the treatment are real, or greater than expected by chance alone. Let's go through a sample experiment.

19-2a Example Design

An experimenter is interested in the difference of finger-tapping speed by the right and left hands. She believes that if a difference is found, it will confirm a theory about hemispheric differences (left versus right) in the brain.

A sample of thirteen subjects (N = 13) is taken from a population of adults. Six subjects tap for fifteen seconds with their right ring finger. Seven subjects tap with their left ring finger. After the number of taps has been recorded, the subjects tap again, but with the opposite hand. Thus each subject appears in each level of the treatment condition, tapping with both the right hand and the left hand.

19-2b Raw Scores

After the data is collected, it is usually arranged in a table like the following:

Example Data

Subject	1	2	3	4	5	6	7	8	9	10	11	12	13
Right Hand	63	68	49	51	54	32	43	48	55	50	62	38	41
Left Hand	65	63	42	31	57	33	38	37	49	51	48	42	37

Note that the two scores for each subject are written in the same column.

19-2c Analysis Step One: Find the Difference Scores

In analysis of crossed designs, first calculate the difference scores for each subject. These scores become the basic unit of analysis. Add a row to the example data table for the difference scores:

Example Data with Difference Scores

Subject	1	2	3	4	5	6	7	8	9	10	11	12	13
Right Hand	63	68	49	51	54	32	43	48	55	50	62	38	41
Left Hand	65	63	42	31	57	33	38	37	49	51	48	42	37
Difference	-2	5	7	20	-3	-1	5	11	6	-1	14	-4	4

The difference scores will be symbolized by D_i to differentiate them from raw scores, which are symbolized by X_i.

19-2d Analysis Step Two: Find the Mean and Standard Deviation of D_i

The next step is to enter the difference scores into the calculator and calculate the mean and standard deviation. For example, for the difference scores in the preceding table (-2, 5, 7, and so on), the mean = $\bar{D}$ = 4.69 and the Standard Deviation = s_D = 7.146.

The mean and standard deviation of the difference scores will be represented by $\bar{D}$ and s_D, respectively.

The Model under the Null Hypothesis

The **null hypothesis** states that there are no effects. That is, if the experiment were conducted with an infinite number of subjects, the average difference between the right and left hand would be zero ($\mu_D = 0$).

If the experiment using thirteen subjects was repeated an infinite number of times assuming the null hypothesis was true, then a model could be created of the means of these experiments. This model would be the sampling distribution of the mean difference scores. The Central Limit Theorem states that the sampling distribution of the mean has a mean equal to the mean of the population model. In this case, the mean would equal 0.0, and a standard error represented by $\sigma_{\bar{D}}$. The standard error of the mean difference scores could be computed by the following formula:

$$\sigma_{\bar{D}} = \frac{\sigma_D}{\sqrt{N}}$$

The only difficulty in this case is that the standard deviation of the population model (σ_D) is not known. It can, however, be estimated using the sample standard deviation, s_D. This estimation adds error to the procedure and requires the use of the t distribution rather than the normal curve. The t distribution will be discussed in greater detail later in this chapter.

19-2e Analysis Step Three: Estimate the Standard Error

The standard error of the mean difference score, $\sigma_{\bar{D}}$, is estimated by $s_{\bar{D}}$, which is calculated by the following formula:

$$s_{\bar{D}} = \frac{s_D}{\sqrt{N}}$$

Using this formula on the example data yields:

$$s_{\bar{D}} = \frac{s_D}{\sqrt{N}} = \frac{7.146}{\sqrt{13}} = 1.982$$

19-2f Analysis Step Four: Find the Degrees of Freedom

Before the exact significance level can be found using the Probability Calculator, an additional parameter, called the degrees of freedom, must be calculated. **Degrees of freedom,** symbolized as df, are basically the number of values that are free to vary. In a sample of N scores, each score can take on a different value, so there are N degrees of freedom. If the mean of the scores is computed and used in a computation, as it is when finding the standard deviation, then one degree of freedom is lost because given that both the mean and N − 1 of the scores are known, the Nth score can always be found. Since this is the situation in the case of a crossed t-test, the degrees of freedom are found using the following formula:

$$df = N - 1$$

Degrees of freedom for crossed t-test.

In this case, it results in the following:

$$df = N - 1 = 13 - 1 = 12$$

Example calculation of degrees of freedom.

19-2g Analysis Step Five: Find the Significance Level

The **exact significance level,** or probability of obtaining a mean difference equal to or greater than the one we obtained given that chance alone was operating, of this statistic may be found using the Probability Calculator. To do so, follow these steps.

1. Select the Two-tailed Sig Level option under t Distribution.

2. Click the t Distribution button.

3. Enter **12** in the df (degrees of freedom) box.
   Enter **0** in the Mu box (mu value for the distribution when null hypothesis is true)
   Enter **1.982** (estimated standard error of the mean difference scores) in the Sigma box.
   Enter **4.69** (mean difference score) in the Score box.

4. Click the right-arrow button to generate the result, which is shown in the following figure:

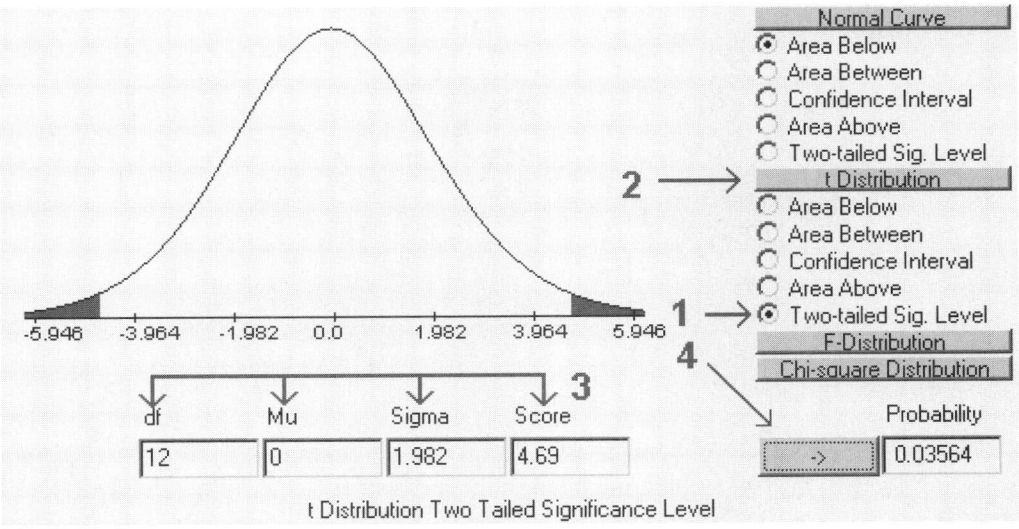

t Distribution Two Tailed Significance Level

Finding the exact significance level in a crossed t-test.

Since the exact significance level equals .036 and is less than the value of alpha (.05), the model under the null hypothesis is rejected and the hypothesis of real effects accepted. The mean difference is said to be statistically significant. In this case the conclusion would be made that the right hand taps faster than the left hand because the mean difference is greater than zero and the number of taps using the left hand was subtracted from the number of taps using the right. If the mean difference were negative, then the left hand would have tapped faster than the right.

19-2h Using SPSS to Compute a Crossed T-Test

The data is entered into an SPSS data file with each subject as a row and the two variables for each subject as columns. In the example data, there would be thirteen rows and two columns, one each for the right and left hand data. The data file looks like the following figure:

	right	left
1	63	65
2	68	63
3	49	42
4	51	31
5	54	57
6	32	33
7	43	38
8	48	37
9	55	49
10	50	51
11	62	48
12	38	42
13	41	37

Entering the data into the SPSS data editor.

What I call a crossed t-test, the SPSS package calls a Paired-Samples T Test. This statistical procedure is accessed by selecting Analyze/Compare Means/Paired-Samples T Test, as the following figure shows:

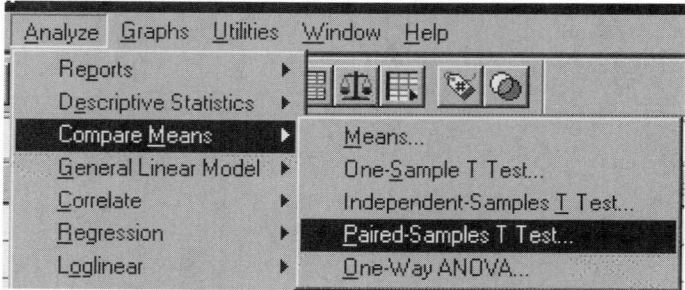

Performing a crossed t-test using SPSS.

You must then tell SPSS which variables are to be compared by double-clicking them. After selecting the two variables to be included in the analysis (see the following figure), click the OK button.

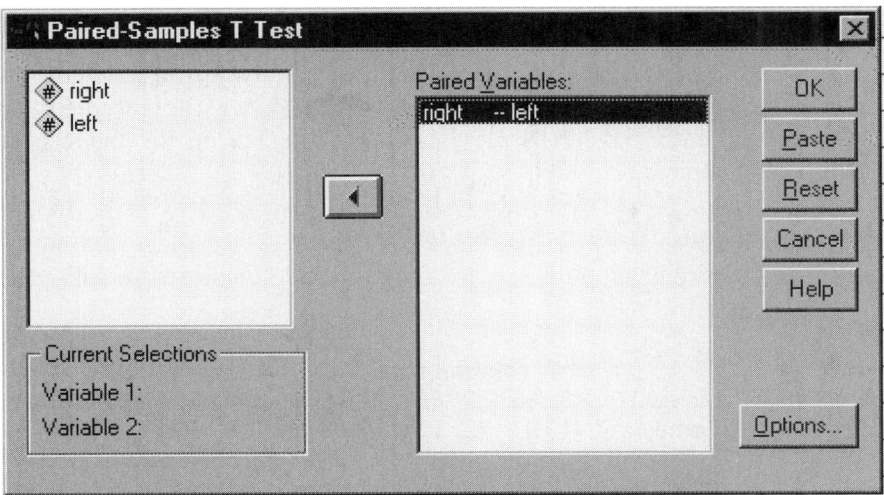

Selecting the variables for a crossed T Test in SPSS.

The first table of SPSS output from this procedure includes the means:

Paired Samples Statistics

		Mean	N	Std. Deviation	Std. Error Mean
Pair 1	RIGHT	50.31	13	10.33	2.87
	LEFT	45.62	13	11.03	3.06

Means and standard deviations are part of the output of the SPSS Paired-Samples T Test.

The second SPSS output table includes the t-test results:

Paired Samples Test

		Paired Differences							
					95% Confidence Interval of the Difference				
		Mean	Std. Deviation	Std. Error Mean	Lower	Upper	t	df	Sig. (2-tailed)
Pair 1	RIGHT - LEFT	4.69	7.15	1.98	.37	9.01	2.368	12	.036

The second table of SPSS output from the Paired-Samples T Test procedure.

Note that the results from using SPSS are within rounding error of the results computed by using the formulas, a hand-held calculator, and the Probability Calculator.

19-3 Nested T-Tests

A nested t-test is the appropriate hypothesis test when there are two groups and different subjects are used in the treatment groups. Two examples of nested t-tests will be presented in this chapter, one detailing the logic underlying the analysis and one illustrating the manner in which the analysis is performed in practice.

19-3a Analysis and Logic Underlying Nested Designs

In a comparison of the finger-tapping speed of males and females, the following data was collected:

Finger Taps

Males	43 56 32 45 36 48
Females	41 63 72 53 68 49 51 59 60

The design is necessarily nested because each subject has only one score and appears in a single treatment condition. Through the marvels of modern medicine it might be possible to treat gender as a crossed design, but finding subjects might be somewhat difficult. The next step is to find the mean and variance of the two groups:

Males	$N_1 = 6$	$\bar{X}_1 = 43.33$	$s_1^2 = 73.47$
Females	$N_2 = 9$	$\bar{X}_2 = 57.33$	$s_2^2 = 95.75$

The critical statistic is the difference between the two means: $\bar{X}_1 - \bar{X}_2$. In the example, it is noted that females, on the average, tapped faster than males. A difference of 43.33–57.33 or -14.00 was observed between the two means. The question addressed by the hypothesis test is whether this difference is large enough to consider the effect to be due to a real difference between males and females, rather than a chance happening (the sample of females just happened to be faster than the sample of males).

The analysis proceeds in a manner similar to all hypothesis tests. An assumption is made that there is no difference in the tapping speed of males and females. The experiment is then carried out an infinite number of times, each time finding the difference between means, creating a model of the world when the null hypothesis is true. The difference between the means that was obtained from the experiment is compared to the difference that would be expected on the basis of the model of no effects. If the difference that was found is unlikely given the model, the model of no effects is rejected and the difference is said to be significant.

In the case of a nested design, the sampling distribution is the difference between means. The sampling distribution consists of an infinite number of differences between means.

The sampling distribution of this statistic is characterized by the parameters $\mu_{\bar{X}_1 - \bar{X}_2}$ and $\sigma_{\bar{X}_1 - \bar{X}_2}$. In the case of $\mu_{\bar{X}_1 - \bar{X}_2}$, if the null hypothesis were true, then the mean of the sampling distribution would be equal to zero (0). The value of $\sigma_{\bar{X}_1 - \bar{X}_2}$ is not known, but may be estimated. In each of the following formulas, the assumption is made that the variances of the population values of σ^2 for each group are similar.

The computational formula for the estimate of the standard error of the difference between means is

$$s_{\bar{x}_1-\bar{x}_2} = \sqrt{\frac{(N_1-1)s_1^2 + (N_2-1)s_2^2}{N_1+N_2-2}\left(\frac{1}{N_1}+\frac{1}{N_2}\right)}$$

The procedure for finding the value of this statistic using a statistical calculator with parentheses is

```
N₁-1    *    s²₁    =    +
(   N₂-1    *    s²₂   )    =    /
N₁+N₂-2    =    *
( N₁ 2nd 1/X + N₂ 2nd 1/X )
   =    SQR(X)
```

If each group has the same number of scores, the preceding formula may be simplified:

if N = N = N, then

$$s_{\bar{x}_1-\bar{x}_2} = \sqrt{\frac{s_1^2+s_2^2}{N}}$$

Because the example problem has different numbers in each group, the longer formula must be used to find the standard error of the differences between means. Computation proceeds as follows:

```
Keys Pressed                         Display

5    *    73.47    =    +              367.35
(   8    *    95.75   )    =    /     1133.35
13    =    *                           87.18
( 6 2nd 1/X + 9 2nd 1/X )                .28
= SQR(X)                                4.92

   where s_{x̄1 -x̄2} = 4.92.
```

The sampling distribution of the differences between means when, in fact, the null hypothesis is true may now be estimated, as illustrated in the following:

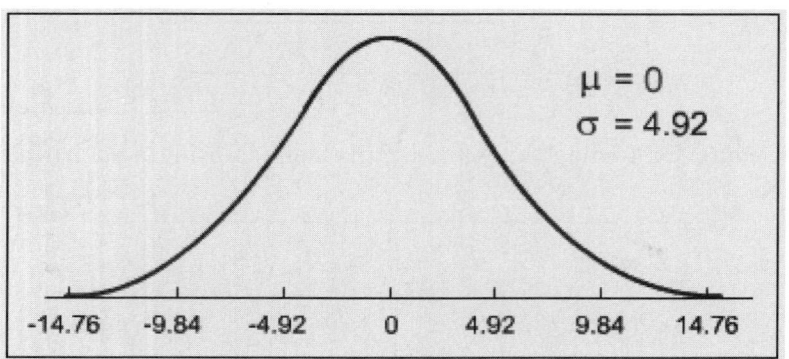

$\mu = 0$

$\sigma = 4.92$

-14.76   -9.84   -4.92   0   4.92   9.84   14.76

Degrees of freedom for a nested t-test.

The degrees of freedom for a nested t-test are N1 + N2 – 2. In this case, there would be 6 + 9 – 2 = 13 degrees of freedom.

$$df = N_1 + N_2 - 2$$

The probability (the exact significance level) of obtaining the observed difference between the means ($\bar{X}_1 - \bar{X}_2 = 43.33 - 57.33 = -14.00$), given that the preceding model is true, is obtained by means of the Probability Calculator. Here are the steps:

1. Select the Two-tailed Sig Level under t Distribution.

2. Click the t Distribution button.

3. Enter a value of **13** in the df (degrees of freedom) box.
    Enter a value of **0** in the Mu box (the value of mu for the distribution when the null hypothesis is true is zero).
    Enter the value **4.92** (the estimated standard error of the difference between the means) in the Sigma box.
    Enter the mean difference score, **-14.00**, in the Value box.

4. Click the right-pointing arrow.

The result, an exact significance level of .014, appears in the following illustration:

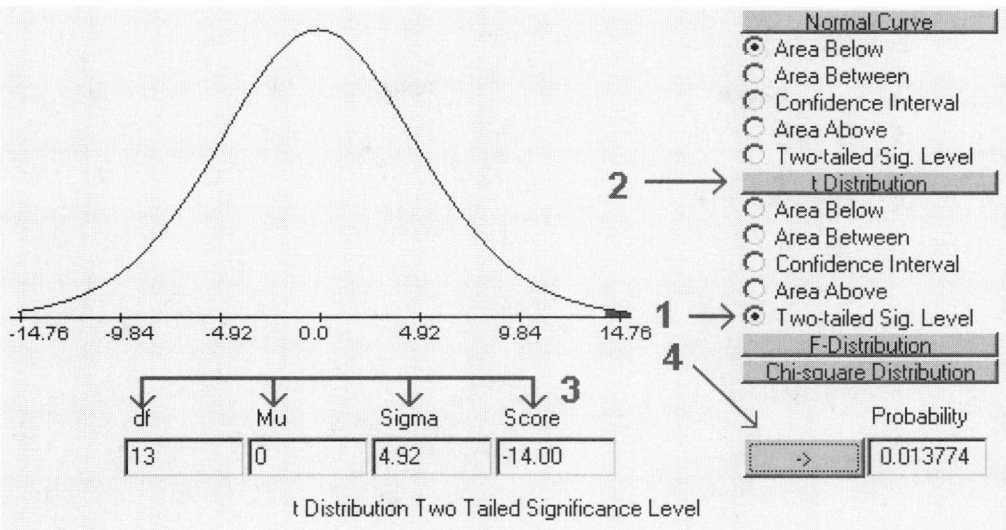

t Distribution Two Tailed Significance Level

Finding the exact significance level in a nested t-test.

Since the value of the exact significance level is less than alpha, the model would be rejected along with the null hypothesis. The alternative hypothesis, that there were real effects, would be accepted. Because the mean for females is larger than the mean for males in this case, the effect would be that females tapped faster than males.

19-3b Computing a Nested T-Test Using SPSS

To compute a **nested t-test** using SPSS, the data for each subject must first be entered with one variable identifying to which group that subject belongs, and a second with the actual score for that subject. In the example data, one variable identifies the gender of the subject and the other identifies the number of finger taps that subject did. A 1 indicates the male subject group and a 2 indicates the female subject group in the gender column of the example data file, as shown in the following figure:

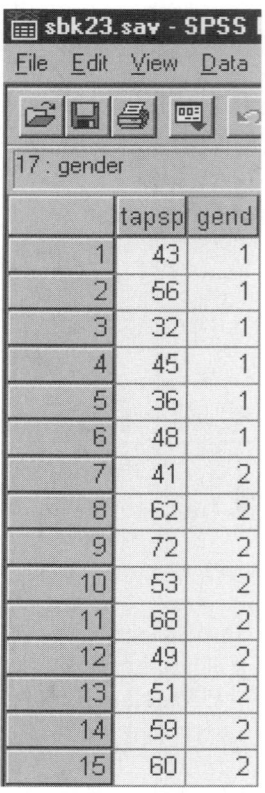

Entering the data in the Data Editor to perform a nested t-test.

The SPSS package does what I call a nested t-test with the Independent-Samples T Test procedure, which is accessed by selecting Analyze/Compare Means/Independent-Samples T Test, as the following figure shows:

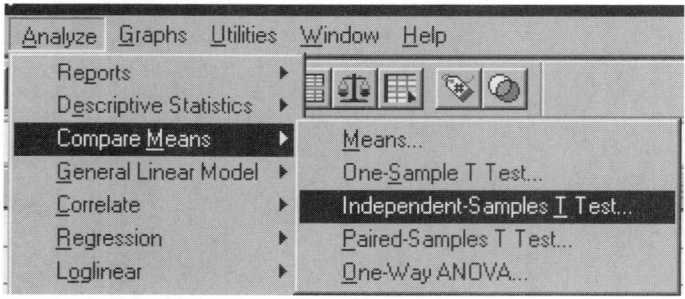

Commands in SPSS to do a nested t-test.

Then you must describe both the dependent (Test) variable and the independent (Grouping) variable to the procedure. The levels of the Grouping Variable are further described by clicking the Define Groups button. You can see example input to the procedure in the following figure:

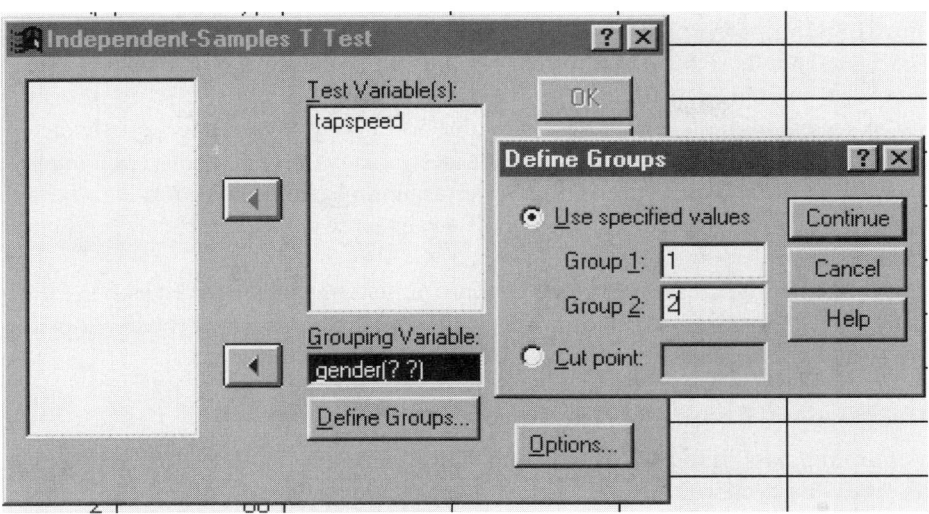

Selecting the variables in SPSS to perform a nested t-test.

Clicking Continue and then OK produces both a table of means, like this one:

Group Statistics

	GENDER	N	Mean	Std. Deviation	Std. Error Mean
TAPSPEED	Male	6	43.33	8.57	3.50
	Female	9	57.33	9.79	3.26

The first table of output using SPSS to compute a nested t-test.

and the results of the nested t-test, like this:

Independent Samples Test

		Levene's Test for Equality of Variances		t-test for Equality of Means						
									95% Confidence Interval of the Difference	
		F	Sig.	t	df	Sig. (2-tailed)	Mean Difference	Std. Error Difference	Lower	Upper
TAP SPE ED	Equal variances assumed	.319	.582	-2.845	13	.014	-14.00	4.92	-24.63	-3.37
	Equal variances not assumed			-2.927	11.865	.013	-14.00	4.78	-24.44	-3.56

The second table of SPSS output for a nested t-test.

The SPSS output produces more detailed results than those presented earlier in the chapter. The Levene's Test for Equality of Variances columns present a test of the assumption that the theoretical variances of the two groups are equal. If this statistic is significant, that is, the value of Sig. is less than .05 (or whatever the value for alpha), then some statisticians argue that a different procedure must be used to compute and test for differences between means. The SPSS output gives results of statistical procedures both assuming equal and not equal variances. In the case of the example analysis, the test for equal variances was not significant, indicating that the first **t-test** would be appropriate.

Note that the procedures described earlier in this chapter produce results within rounding error of those assuming equal variances. When in doubt, you should probably opt for the more conservative (less likely to find results) t-test assuming unequal variances.

19-3c An Example without Explanations

In an early morning class, one-half the students are given coffee with caffeine and one-half coffee without caffeine. The number of times each student yawns during the lecture is recorded with the following results:

Number of Yawns with and without Caffeine

Caffeine	3	5	0	12	7	2	5	8
No Caffeine	8	9	17	10	4	12	16	11

Because each student participated in one, and only one, treatment condition, the experiment has subjects nested within treatments.

Step One

Caffeine	$N_1 = 8$	$\bar{X}_1 = 5.25$	$s_1^2 = 14.21$
No Caffeine	$N_2 = 8$	$\bar{X}_2 = 10.875$	$s_2^2 = 17.84$

Example means and variances for a nested t-test.

Step Two

$$s_{\bar{X}_1 - \bar{X}_2} = \sqrt{\frac{s_1^2 + s_2^2}{N}} = \sqrt{\frac{14.21 + 17.84}{8}} = 2.00$$

Estimating the standard error of the differences between means.

Step Three

The degrees of freedom are N1 + N2 − 2; in this case, 8 + 8 − 2 = 14.

Step Four

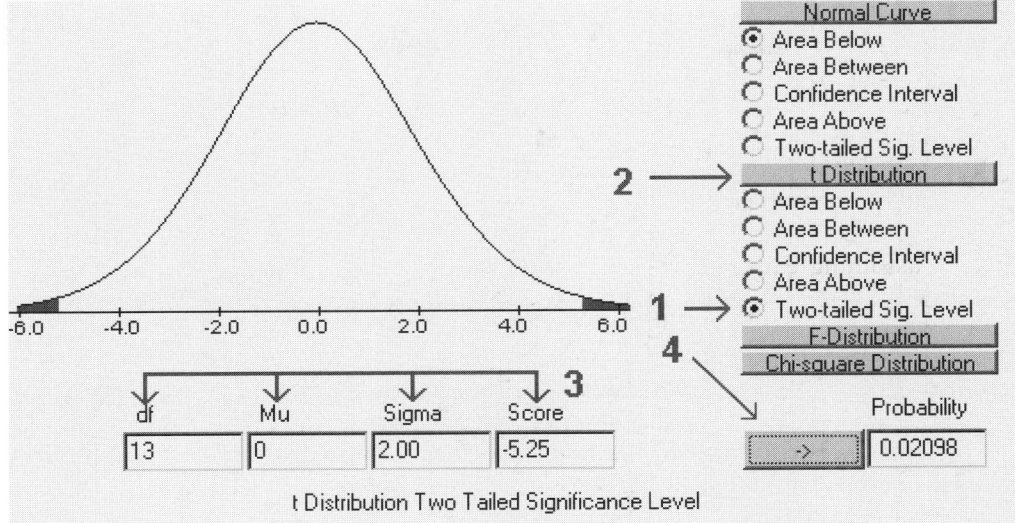

t Distribution Two Tailed Significance Level

Finding the exact significance level in a nested t-test.

Step Five

The null hypothesis is rejected because the value for the exact significance level is less than the value of alpha. In this case, it is possible to say that caffeine had a real effect.

19-4 The T Distribution

The **t distribution** is a theoretical probability distribution. It is symmetrical, bell-shaped, and similar to the normal curve. It differs from the normal curve, however, in that it has an additional parameter, called degrees of freedom, that changes its shape.

19-4a Degrees of Freedom

Degrees of freedom, usually symbolized by df, is a parameter of the t distribution that can be any real number greater than zero (0.0). Setting the value of df in addition to the values of mu and sigma defines a particular member of the family of t distributions. A member of the family of t distributions with a smaller df has more area in the tails of the distribution than one with a larger df.

The effect of df on the t distribution is illustrated in the three t distributions in the following figure:

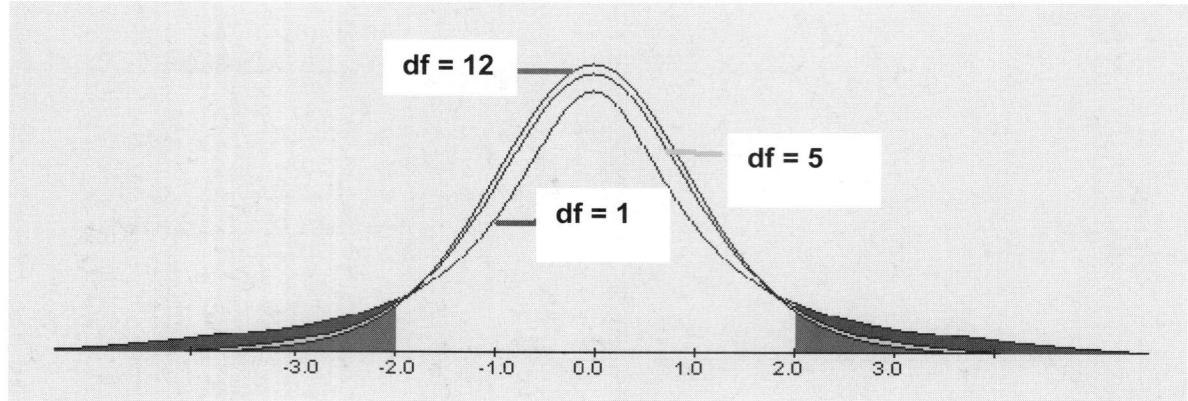

Changes in the t distribution as a function of the degrees of freedom.

Note that the smaller the df, the flatter the shape of the distribution, resulting in greater area in the tails of the distribution.

19-4b Relationship to the Normal Curve

The astute reader will no doubt observe that the t distribution looks very similar to the normal curve. As the df increase, the t distribution approaches the normal distribution with similar values for mu and sigma. The normal curve is a special case of the t distribution when df = ∞. For practical purposes, the t distribution approaches the normal distribution relatively quickly, such that when df = 30 the two are almost identical.

19-4c Testing Hypotheses with the T Distribution

The t distribution rather than the normal curve is used to test hypotheses when the value of sigma must be estimated rather than being known. Estimating the value of sigma adds additional uncertainty or error to the hypothesis testing procedure. Statisticians attempt to correct for this error by making the hypothesis test more conservative, that is, less likely to find results, and this is accomplished by using the t distribution rather than the normal curve.

For example, the exact significance level when mu = 0, sigma = 2, and the value = 3.5 is .080 for the normal curve, .106 for the t distribution with df = 12, .141 for the t distribution with df = 5, and .330 for the t distribution with df = 1. Because the decision to reject the null hypothesis is made by comparing the exact significance level to alpha, the smaller the degrees of freedom, the less likely we are to reject the null hypothesis. Thus the hypothesis test becomes more conservative the smaller the degrees of freedom. Basically, the smaller the sample size, the smaller the degrees of freedom, the greater the likelihood of error in estimating the value of sigma, the greater the correction that is made by using the t distribution rather than the normal curve.

19-5 One- and Two-Tailed T-Tests

A choice of a one- or two-tailed t-test determines how the exact significance level is computed. The **one-tailed t-test** is performed if the results are interesting only if they turn out in a particular direction. The **two-tailed t-test** is performed if the results would be interesting in either direction. The choice of a one- or two-tailed t-test must be made in the design stage of the study and affects the hypothesis testing procedure in a number of different ways.

19-5a Two-Tailed T-Tests

A two-tailed t-test computes the exact significance level by summing the areas in two tails under the t distribution. If the score were greater than mu, the areas summed would be the area under the curve above the score and the area below a mirror image of the score under the same distribution. In the following example, df = 5, mu = 0, sigma = 3.47, and the score = 4.11. The exact significance level is computed by summing the areas above 4.11 and below -4.11.

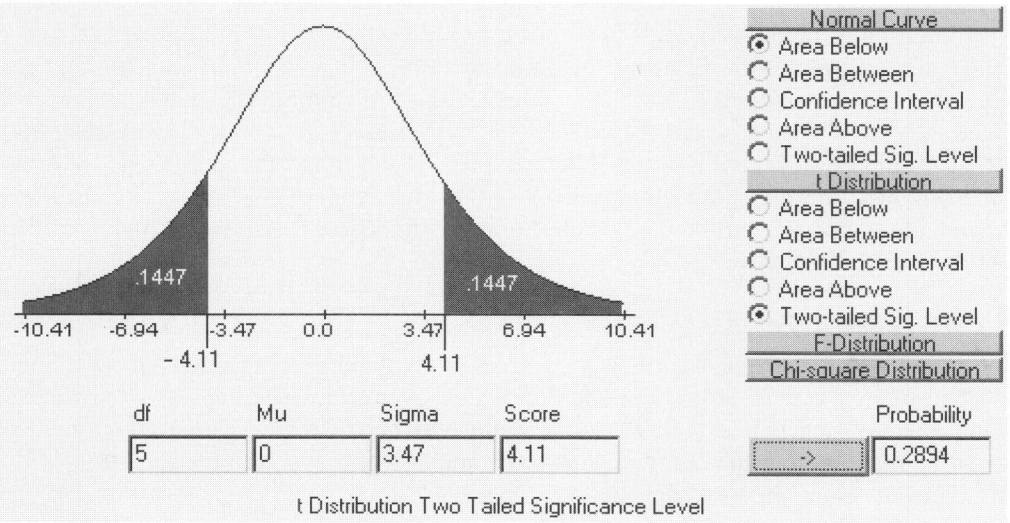

Finding the exact significance level in a two-tailed t-test with a positive score.

If the score were less than mu, the areas summed would be the area below the score plus the area above a mirror image of the score. In the following example, df = 5, mu = 0, sigma = 3.47, and the score = -2.39. The exact significance level is computed by summing the areas below -2.39 and above 2.39.

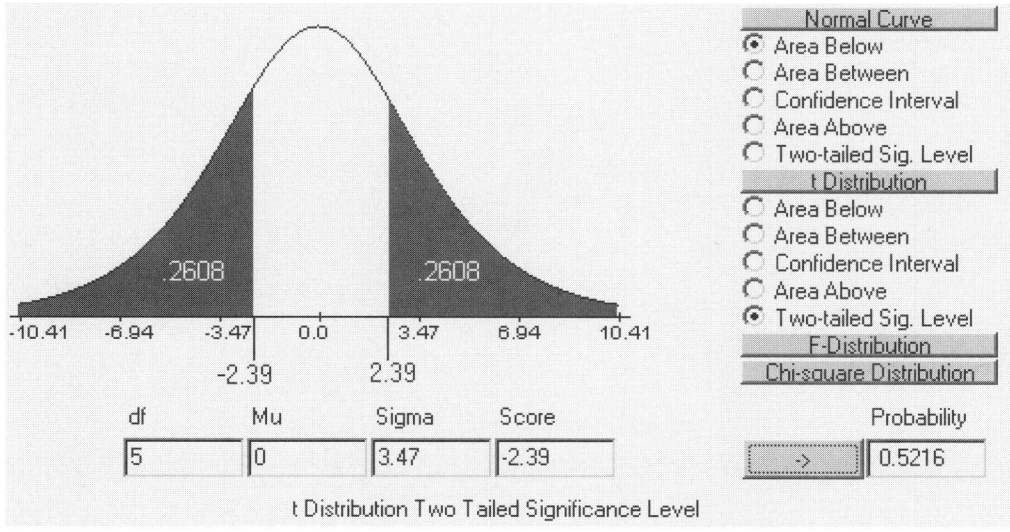

Finding the exact significance level in a two-tailed t-test with a negative score.

In either case, because of the symmetry of the t distribution the exact significance level will be twice the area either above or below the score, whichever is less.

When doing a two-tailed t-test, the null hypothesis is a particular value, and there are two alternative hypotheses, one positive and one negative. When the null hypothesis is rejected, one of the alternative hypotheses is accepted and the other rejected, depending upon the direction of the results.

For example, consider the earlier example study comparing the finger-tapping speeds of males and females. If a two-tailed nested t-test were done comparing the mean finger-tapping speeds of males and females, the null hypothesis would be that males and females tapped equally fast. One alternative hypothesis would be that males tapped faster than females and the other would be that females tapped faster than males. If the null hypothesis that males and females tapped equally fast was rejected based on the hypothesis test, then one of the alternative hypotheses would be accepted and the other rejected. If the mean for males was larger than the mean for females, the alternative hypothesis that males tapped faster than females would be accepted. If the opposite was the case, then the alternative hypothesis that females tapped faster than males would be accepted.

19-5b One-Tailed T-Tests

In a one-tailed t-test, the exact significance level is calculated by finding the area under the t distribution either above or below the score. Two different one-tailed t-tests exist: one for area above and one for area below. The decision about whether to find the area above or below the score must be made before the study is conducted and corresponds with the direction of the score value if the results come out as expected.

A One-Tailed T-Test in the Positive Direction

In a one-tailed t-test in the positive direction, the exact significance level is computed by finding the area above the score value. In the following example, df = 5, mu = 0, sigma = 3.47, and the score = 4.11. The exact significance level is computed by finding the area above 4.11.

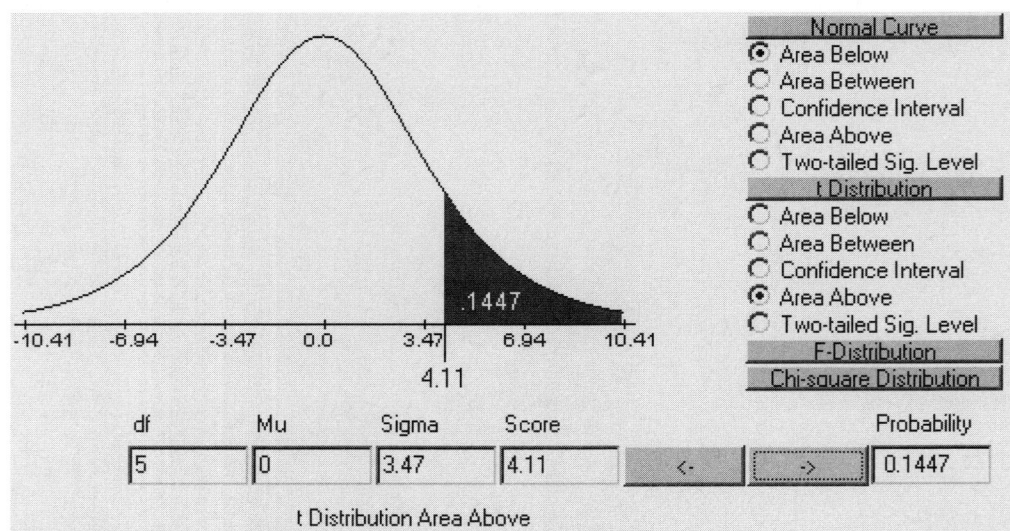

Finding the exact significance level in a one-tailed t-test with a positive score.

In a second example, this time with a negative score, df = 5, mu = 0, sigma = 3.47, and the score = -2.39. The exact significance level is computed by finding the area above -2.39. In this case, the results of the study were opposite the prediction and the exact significance level is very large at .7392.

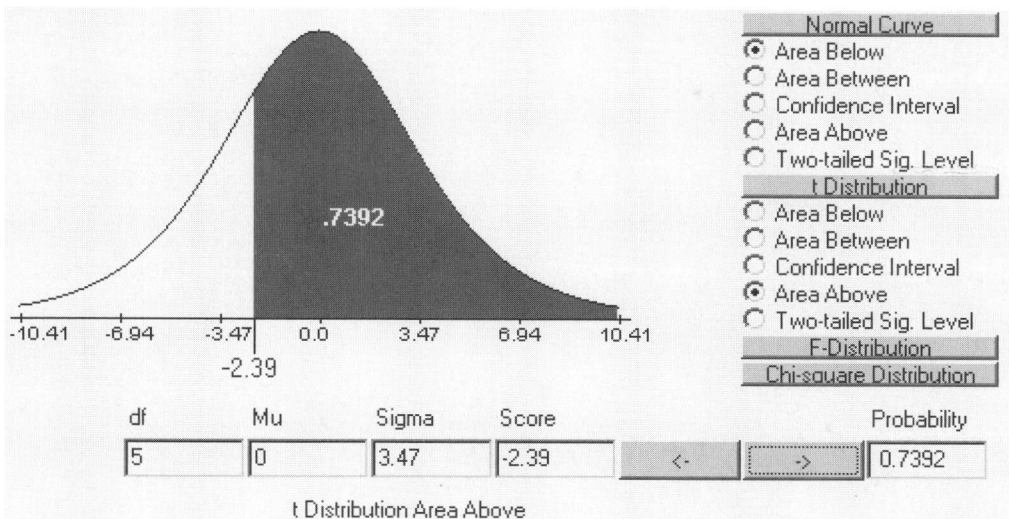

t Distribution Area Above

Finding the exact significance level in a one-tailed t-test with a negative score.

A One-Tailed T-Test in the Negative Direction

In a one-tailed t-test in the negative direction, the exact significance level is computed by finding the area below the score value. In the following example, df = 5, mu = 0, sigma = 3.47, and the score = 4.11. The exact significance level is computed by finding the area below 4.11. In this case, because the score value is opposite the prediction, the exact significance level is large: .8553.

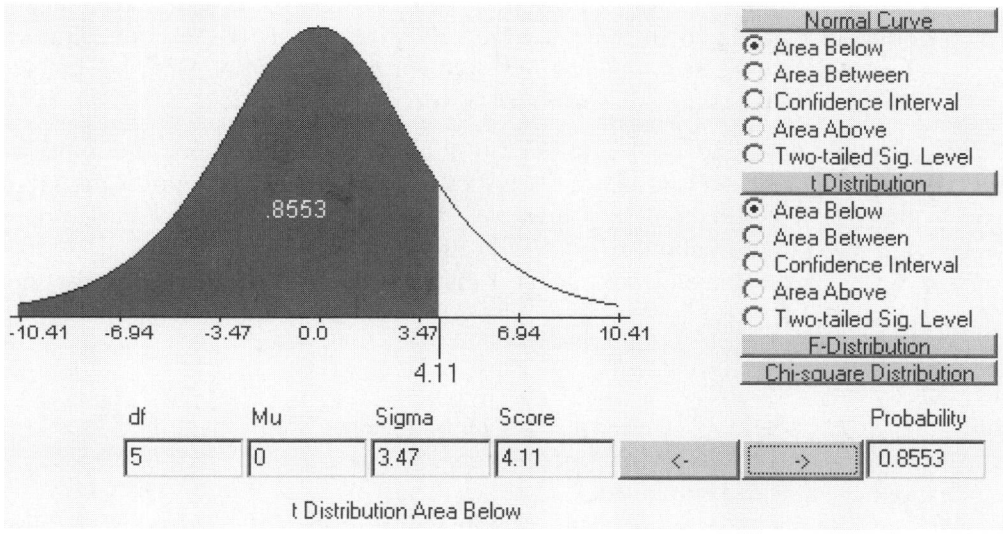

t Distribution Area Below

Finding the exact significance level in a one-tailed t-test with a positive score.

In a second example, this time with a negative score, df = 5, mu = 0, sigma = 3.47, and the score = -2.39. The exact significance level is computed by finding the area below -2.39. In this case, the results of the study were congruent with the prediction and the exact significance level is .2608.

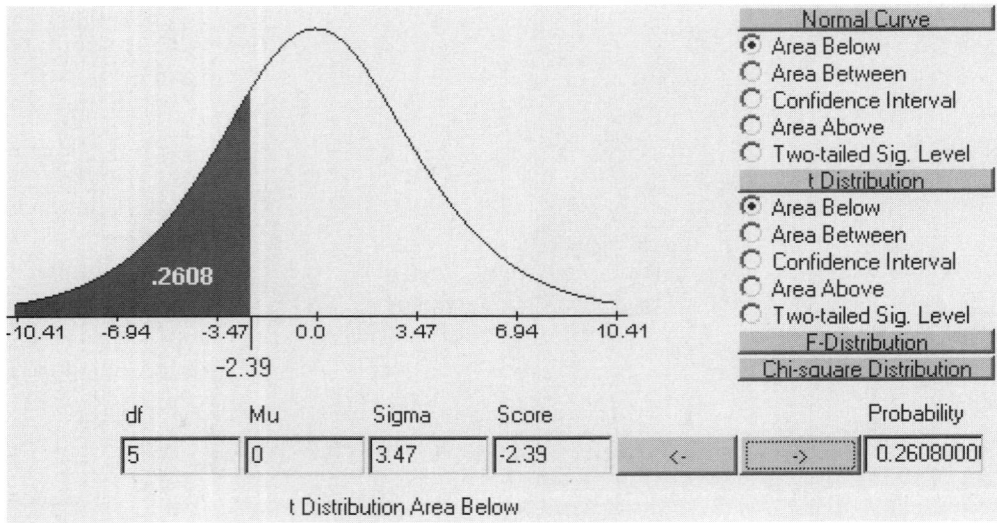

Finding the exact significance level in a one-tailed t-test with a negative score.

19-5c Comparison of One- and Two-Tailed T-Tests

1. If alpha = .05, df = 12, mu = 0, sigma = 3.47, and the score = 7.9, then significance would be found in the two-tailed and the *positive* one-tailed t-tests. The one-tailed t-test in the negative direction would not be significant, because the prediction chose the wrong direction. This is the danger of a one-tailed t-test.

2. If alpha = .05, df = 12, mu = 0, sigma = 3.47, and the score = -6.5, then significance would only be found in the *negative* one-tailed t-test. If the correct direction is selected, it can be seen that the null hypothesis is more likely to be rejected. The significance test is said to have greater power in this case.

The selection of a one or two-tailed t-test must be made before the experiment is performed. It is not "cricket" to find that the score of -6.5 would only be significant with a one-tailed test in the negative direction, and then say, "I really meant to do a one-tailed t-test." Because reviewers of articles submitted for publication are sometimes suspicious when a one-tailed t-test is done, the recommendation is that if there is any doubt, a two-tailed test should be done.

Summary

When only two levels or groups are used in an experiment or study, hypothesis testing about means can be done using a t-test. A different procedure will be used for the t-test depending upon the type of experimental design employed.

When each subject sees each level of treatment, a crossed design results. In a crossed design the effects are found as mean difference scores. The difference scores between the two treatment levels are first calculated for each subject, followed by the mean and standard deviation of the difference scores. The standard error of the mean difference scores is estimated and the exact significance level found by comparing the mean difference score found in the study with a theoretical distribution of mean difference scores given there were no effects. The exact significance level is then compared with the value of alpha that was set before the study was conducted. If the exact significance level is less than alpha, the null hypothesis is rejected and the hypothesis of real effects is accepted. In the opposite case, where the exact significance level is greater than alpha, the null hypothesis is retained with no conclusion about the reality of effects.

When the experimental design has subjects appearing in one, and only one, of two treatment conditions, a nested t-test is the appropriate analysis. In a nested design the effects are found using the differences between the means. The results of the study are compared with a model of the distribution of differences between sample means under the null hypothesis, resulting in an exact significance level. As in all hypothesis tests, if the model is unlikely given the data, the model and the null hypothesis are rejected and the alternative accepted.

The t distribution is used to create the model under the null hypothesis rather than the normal distribution because error is added to the hypothesis testing procedure when the standard errors are estimated rather than known. The t distribution is very similar to the normal distribution, except an additional parameter, called the degrees of freedom, is needed to describe a particular member of this family of distributions. The larger the degrees of freedom, the more similar the t distribution is to the normal distribution and the less correction for error.

The statistician has the choice of using either a single tail or both tails when doing a t-test. In a nondirectional or two-tailed t-test, alpha is divided in half, and each half is placed in one tail of the distribution. The effects may be significant in either direction. In a directional t-test, alpha is placed in a single tail of the distribution and is only significant if the results fall in that tail. Results are more likely to be significant in a directional test, but because it is difficult to verify that the statistician decided before the study to use a one-tailed t-test, the two-tailed t-test is most often used.

20

Errors in Hypothesis Testing

Key Terms

Type I error
Type II error

20-1 Will Mathematics Machines Raise Math Scores?

A superintendent in a medium-size school has a problem. The mathematics scores on nationally standardized achievement tests (such as the SAT and ACT) of the students attending her school are lower than the national average. The school board members, who don't care whether the football or basketball teams win or not, are greatly concerned about this deficiency. The superintendent fears that if the situation is not corrected, she will lose her job before long.

As the superintendent sat in her office wondering what to do, a salesperson approached with a briefcase and a sales pitch. The salesperson had heard about the problem of the mathematics scores and was prepared to offer the superintendent a "deal she couldn't refuse." The product was teaching machines to teach mathematics, guaranteed to increase the mathematics scores of the students. In addition, the machines never take breaks or demand a pay increase.

The superintendent agreed that the machines might work, but was concerned about the cost. The salesperson finally wrote some figures. Since there were about 1,000 students in the school and one machine was needed for every ten students, the school would need about one hundred machines. At a cost of $10,000 per machine, the total cost to the school would be about $1,000,000. As the superintendent picked herself up off the floor, she said she would consider the offer, but didn't think the school board would go for such a big expenditure without prior evidence that the machines actually worked. Besides, how did she know that the company that manufactures the machines might not go bankrupt in the next year, meaning the school would be stuck with a million dollar's worth of useless electronic junk?

The salesperson was prepared, making an offer to lease ten machines for testing purposes to the school for one year at a cost of $500 each. At the end of a year, the superintendent would make a decision about the effectiveness of the machines. If they worked, she would pitch them to the school board; if not, then she would return the machines with no further obligation.

An experimental design was agreed upon. One hundred students would be randomly selected from the student population and would be taught using the machines for one year. At the end of the year, the mean mathematics scores of those students would be compared to the mean scores of the students who did not use the machine. If the means were different enough, the machines would be purchased. (The astute statistics student will recognize this as a nested t-test.)

In order to help decide how different the two means would have to be in order to buy the machines, the superintendent did a theoretical analysis of the decision process. Her analysis is presented in the following decision box:

Decision Boxes in Hypothesis Testing		
	"Real World"	
Decision	**The machines do NOT work.**	**The machines work.**
Buy the machines. Decide the machines work.	(1) Type I ERROR probability = α	(4) CORRECT probability = $1 - \beta$ "power"
Do not buy the machines. Decide that the machines do not work	(2) CORRECT probability = $1 - \alpha$	(3) Type II ERROR probability = β

The decision box has the choices for the decision that the superintendent must make in the left column. For simplicity's sake, only two possibilities are permitted: either buy all the machines or buy none of the machines. The other two columns titles represent "the state of the real world." The state of the real world can never be truly known, because if it were known whether or not the machines worked, there would be no point in doing the experiment. The four "Real World" cells represent various places one could be, depending upon the state of the world and the decision made. Each cell will be discussed in turn.

1. **Buying the machines when they do not work.**

    This is called a **Type I error** and in this case is very costly ($1,000,000). The probability of this type of error is α, also called the significance level, and is directly controlled by the experimenter. Before the experiment begins, the experimenter directly sets the value of α. In this example, the value of α would be set low, lower than the usual value of .05—perhaps as low as .0001, which means that one time out of 10,000 the experimenter would buy the machines when they didn't work.

2. **Not buying the machines when they really didn't work.**

    This is a correct decision, made with probability $1 - \alpha$ when, in fact, the teaching machines don't work and the machines are not purchased.

    The relationship between the probabilities in these two decision boxes can be illustrated (see the following figures) using the sampling distribution when the null hypothesis is true. The decision point is set by α, the area in the tail or tails of the distribution. Setting α smaller moves the decision point further into the tails of the distribution, as you can see in the second distribution.

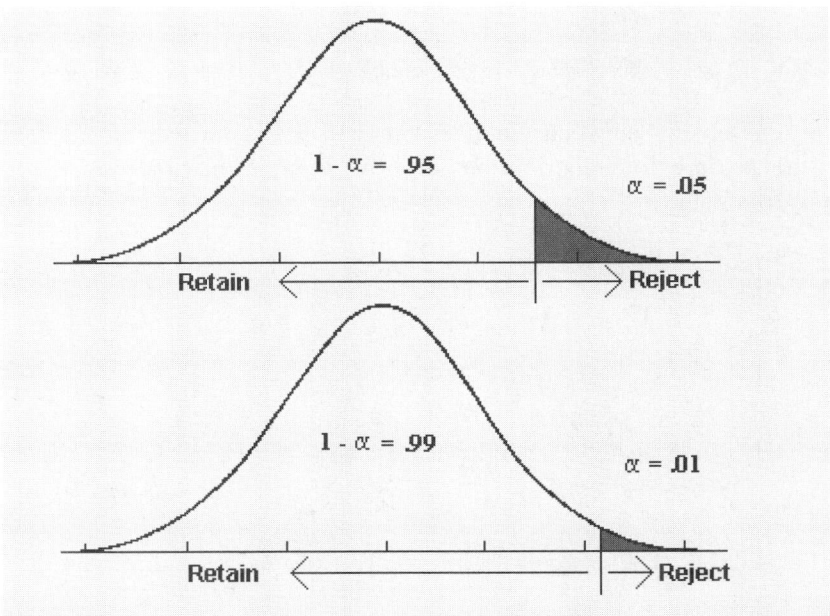

Changing the alpha level changes the critical value of the statistic.

3. **Not buying the machines when they really work.**

    This is called a **Type II error** and is made with probability β. The value of β is not directly set by the experimenter, but is a function of a number of factors, including the size of α, the size of the effect, the size of the sample, and the variance of the original distribution. The value of β is inversely related to the value of α: the smaller the value of α, the larger the value of β. It can now be seen that setting the value of α to a small value was not done without cost, as the value of β is increased.

4. **Buying the machines when they really work.**

This is the cell where the experimenter would usually like to be. The probability of making this correct decision is $1 - \beta$ and is given the name "power." Because α was set low, β would be high, and as a result $1 - \beta$ would be low. Thus it would be unlikely that the superintendent would buy the machines, even if they did work.

The relationship between the probability of a Type II error (β) and power ($1 - \beta$) is illustrated in the following sampling distribution when there actually was an effect:

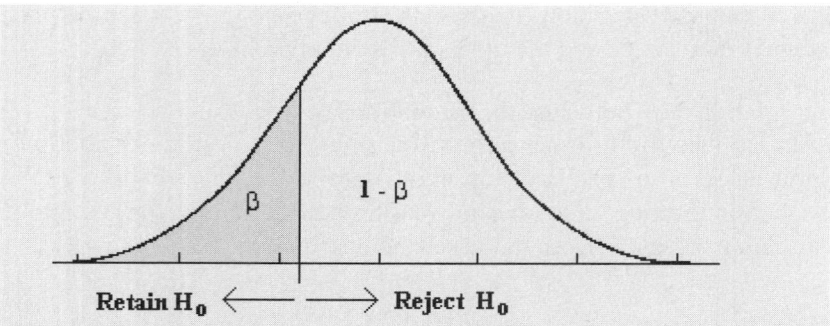

The probability of a Type II error shown on a distribution when the effects are real.

The relationship between the size of α and β can be seen in the following illustration combining the two previous distributions into overlapping distributions, the top graph with $\alpha = .05$ and the bottom with $\alpha = .01$.

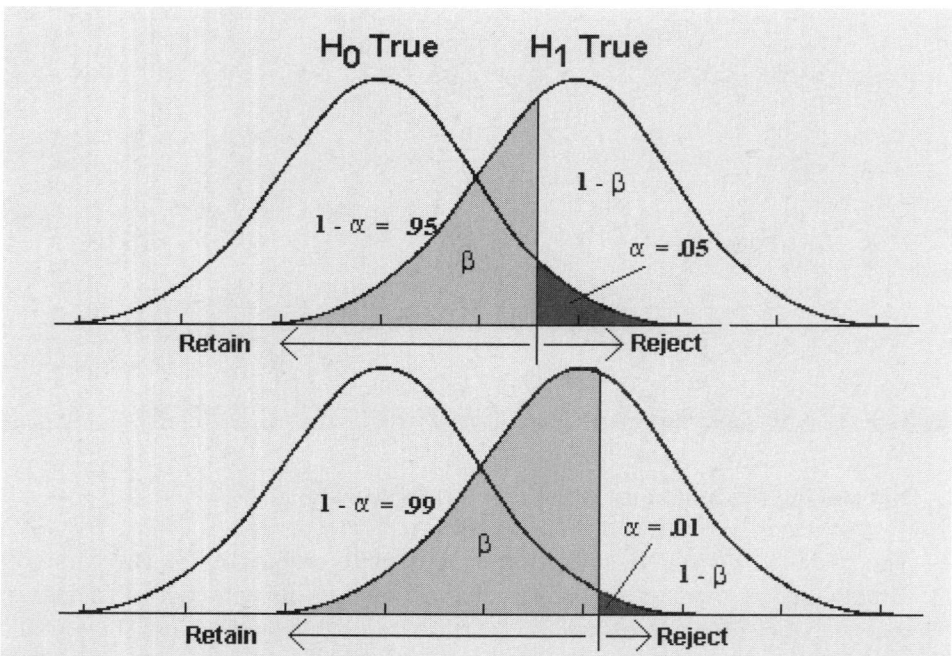

Decreasing the probability of a Type I error (alpha) increases the probability of a Type II error (beta).

The size of the effect is the difference between the center points (μ) of the two distributions. As the size of the effect is increased, the size of beta is decreased. The following distributions show this change:

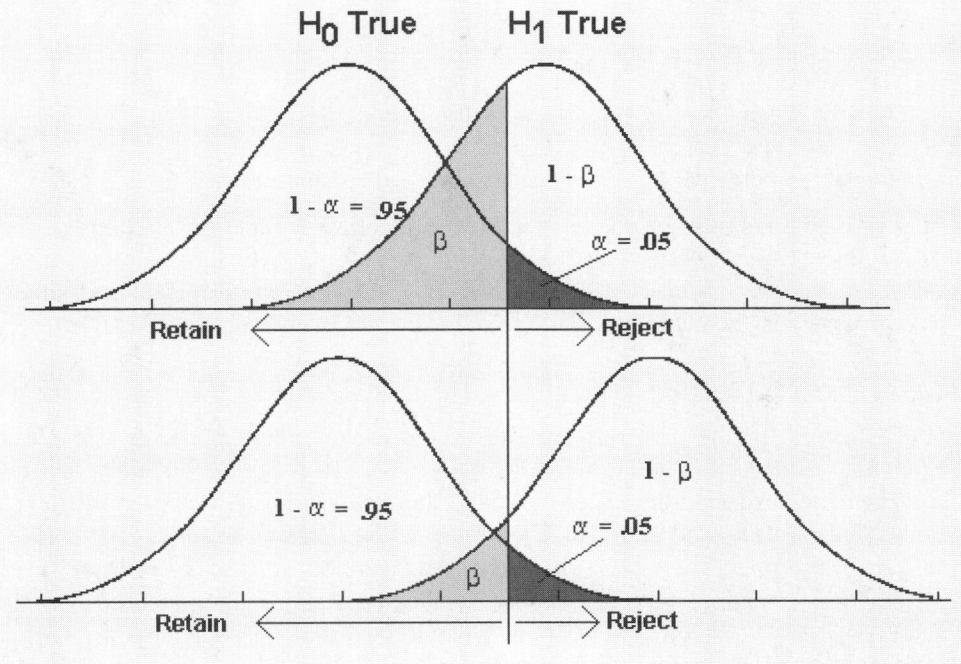

Keeping the value of alpha constant, increasing the size of the effects decreases beta.

When the error variance of the scores is decreased and everything else remains constant, the probability of a Type II error is decreased, as illustrated here:

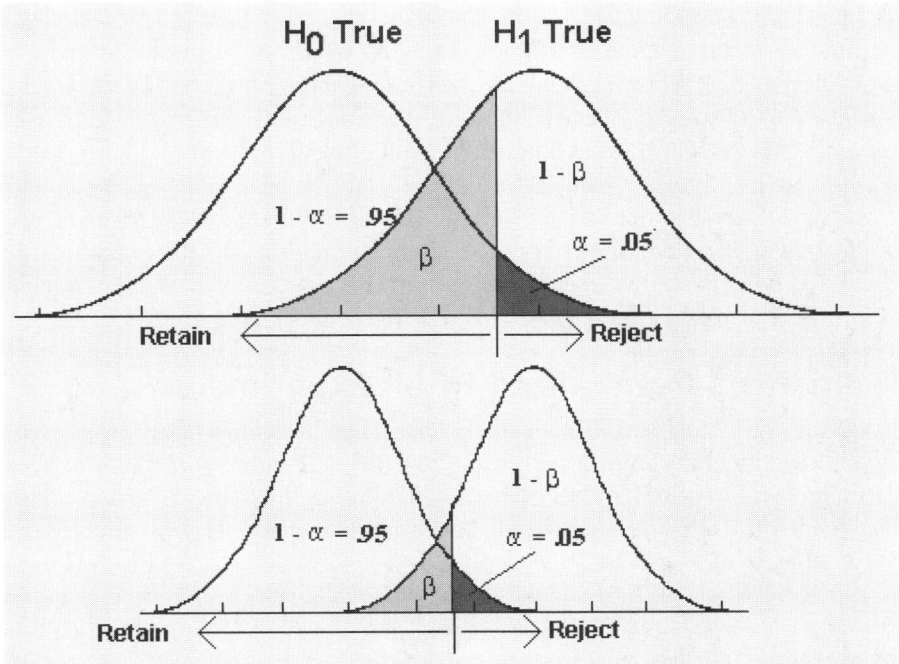

Keeping the size of effects and alpha constant, decreasing the error variance will decrease the size of beta.

The relationship between alpha, size of effects, size of sample (N), size of error, and beta can be listed as follows:

1. The size of beta decreases as the size of error decreases.

2. The size of beta decreases as the size of the sample increases.

3. The size of beta decreases as the size of alpha increases.

4. The size of beta decreases as the size of the effects increases.

The size of the increase or decrease in beta is a complex function of changes in all of the other values. For example, changes in the size of the sample may have either small or large effects on beta, depending upon the other values. If a large treatment effect and small error is present in the experiment, then changes in the sample size are going to have a small effect.

20-2 A Second Chance

As might be expected, in the previous situation the superintendent chose not to purchase the teaching machines, because she had essentially stacked the deck against deciding that there were any effects. When she described the experiment and the result to the salesperson the next year, the salesperson listened carefully and understood the reason why α had been set so low.

The salesperson had a new offer to make, however. Because of an advance in microchip technology, the entire teaching machine had been placed on a single integrated circuit. As a result the price had dropped to $500 a machine. Now it would cost the superintendent a total of $50,000 to purchase the machines, a sum that is quite reasonable.

The analysis of the probabilities of the two types of errors revealed that the cost of a Type I error—buying the machines when they really don't work ($50,000)—is small when compared to the loss encountered in a Type II error—when the machines are not purchased when in fact they do work—although it is difficult to put into dollars the cost of the students not learning to their highest potential.

In any case, the superintendent would probably set the value of α to a fairly large value (.10 perhaps) relative to the standard value of .05. This would have the effect of decreasing the value of β and increasing the power $(1 - \beta)$ of the experiment. Thus the decision to buy the machines would be made more often if, in fact, the machines worked. The experiment was repeated the next year under the same conditions as the previous year, except that the size of α was set to .10.

The results of the significance test indicated that the means were significantly different, the null hypothesis was rejected, and a decision about the reality of effects made. The machines were purchased, the salesperson earned a commission, the math scores of the students increased, and everyone lived happily ever after.

20-3 The Analysis Generalized to All Experiments

The analysis of the reality of the effects of the teaching machines may be generalized to all significance tests. Rather than buying or not buying the machines, you reject or retain the null hypothesis. In the "real world," rather than the machines working or not working, the null hypothesis is true or false. The following decision box presents the choices representing significance tests in general:

Decision Boxes in Hypothesis Testing

Decision	"Real World"	
	Null True **Alternative False** *No Effects*	**Null False** **Alternative True** *Real Effects*
Reject null Accept alternative Decide there are real effects.	Type I ERROR probability = α	CORRECT probability = $1 - \beta$ "power"
Retain null Retain alternative Decide that no effects were discovered.	CORRECT probability = $1 - \alpha$	Type II ERROR probability = β

Summary

When doing a hypothesis test, two types of decision errors are possible. The first, called a Type I error, occurs when the null hypothesis is rejected when in fact it is true. The probability of a Type I error is called alpha and symbolized by α. Alpha is directly set by the researcher with a generally accepted default value of .05. The second type of error is called a Type II error and occurs when the researcher retains the null hypothesis when in fact it is false. The probability of a Type II error is called beta and symbolized by β. The value of beta is indirectly set by the researcher and depends upon four values—alpha, effect size, sample size, and error variance. In general, while the size of alpha is known, the size of beta can only be imprecisely estimated.

It is not difficult to conceive of situations where the default value of .05 for alpha should be abandoned for values that take into account the relative costs of each type of error. Since the probabilities of alpha and beta are inversely related, if the cost of a Type I error is high relative to the cost of a Type II error, then the probability of a Type I error (α) should be set relatively low. If the cost of a Type I error is low relative to the cost of a Type II error, then the probability of a Type I error (α) should be set relatively high.

Chapter

21

Analysis of Variance (ANOVA)

Key Terms

analysis of variance
ANOVA
F-distribution

F-ratio
Mean Squares Between (MS_B)
Mean Squares Within (MS_W)

multiple t-tests

21-1 How Effective Are Various Methods of Therapy?

Multiple comparisons using t-tests is not the analysis of choice. An example can illustrate why.

Suppose a researcher performs a study on the effectiveness of various methods of individual therapy. The methods used are Reality Therapy, Behavior Therapy, Psychoanalysis, Gestalt Therapy, and, of course, a control group. Six patients are randomly assigned to each group. At the conclusion of the study, changes in self-concept are found for each patient. The purpose of the study was to determine if one method was more effective than the other methods in improving patients' self-concept.

At the conclusion of the experiment the researcher creates a data file in SPSS in the following manner:

	therapy	selfcon
2	1	23
3	1	31
4	1	22
5	1	36
6	1	41
7	2	34
8	2	56
9	2	45
10	2	64

The SPSS data file for therapy effectiveness.

The researcher wants to compare the means of the groups to decide about the effectiveness of the therapy.

One method of performing this analysis is by doing all possible t-tests, called **multiple t-tests.** That is, Reality Therapy is first compared with Behavior Therapy, then Psychoanalysis, then Gestalt Therapy, and then the Control Group. Behavior Therapy is then individually compared with the last three groups, and so on. Using this procedure, ten different t-tests would be performed. Therein lies the difficulty with multiple t-tests.

First, because the number of t-tests increases geometrically as a function of the number of groups, analysis becomes cognitively difficult somewhere in the neighborhood of seven different tests. An **analysis of variance** organizes and directs the analysis, allowing easier interpretation of results.

Second, by doing a greater number of analyses, the probability of committing at least one Type I error somewhere in the analysis greatly increases. The probability of committing at least one Type I error in an analysis is called the experiment-wise error rate. The researcher may want to perform a fewer number of hypothesis tests in order to reduce the experiment-wise error rate. The ANOVA procedure performs this function.

In this case, the correct analysis in SPSS is a one-way Analysis of Variance or **ANOVA.** Begin the procedure by selecting Statistics/Compare Means/One-Way ANOVA, as the following figure illustrates.

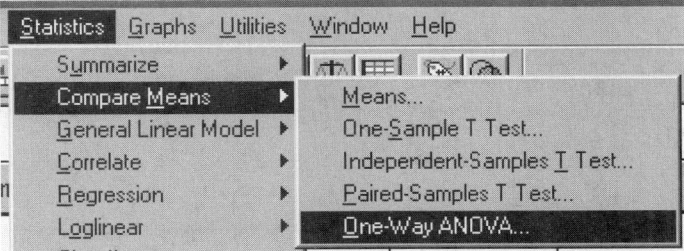

The first step in performing an ANOVA using SPSS.

Then select the variables and options, as shown in this figure:

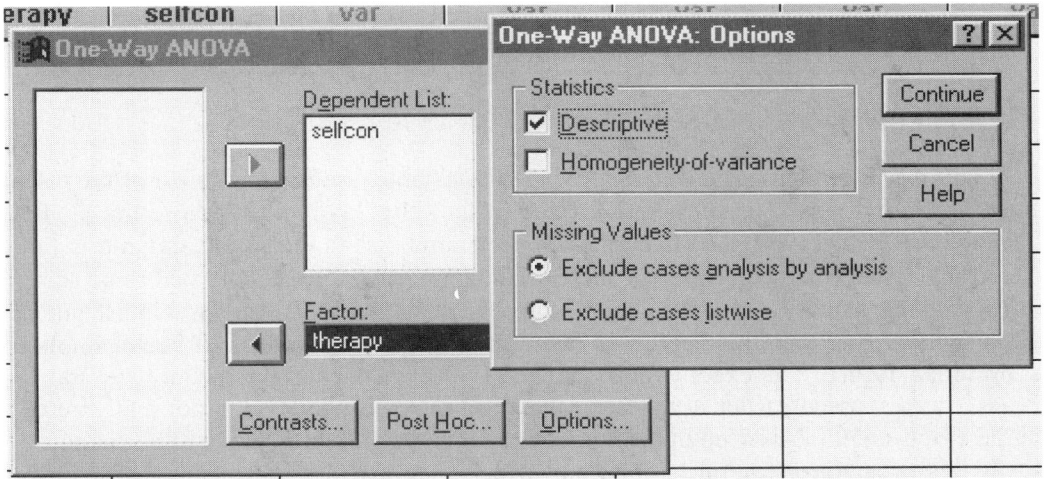

Selecting the variables and options in the SPSS One-Way ANOVA command.

21-2 The Bottom Line: Results and Interpretation of ANOVA

When you check the Descriptive box in the Statistics section of the One-Way ANOVA: Options dialog box (shown in the preceding figure), the result is a table of means and standard deviations, such as the following:

Descriptives

THERAPY	N	Mean	Std. Deviation	Std. Error	95% Confidence Interval for Mean Lower Bound	95% Confidence Interval for Mean Upper Bound
Reality	6	27.50	10.56	4.31	16.42	38.58
Behavior	6	55.83	13.96	5.70	41.18	70.49
Psychoanalysis	6	21.33	7.39	3.02	13.57	29.09
Gestalt	6	49.17	8.01	3.27	40.76	57.57
Control	6	19.00	4.86	1.98	13.90	24.10
Total	30	34.57	17.66	3.22	27.97	41.16

SPSS descriptive statistics output for the ANOVA command.

The results of the ANOVA are presented in an ANOVA table, which has columns labeled Sum of Squares (sometimes referred to as SS), df (degrees of freedom), Mean Square (sometimes referred to as MS), F (for F-ratio), and Sig. The only column that is critical for interpretation is the last (Sig.)! The others are used mainly for intermediate computational purposes. Here's an example of an ANOVA table:

ANOVA

		Sum of Squares	df	Mean Square	F	Sig.
Self-Concept	Between Groups	6796.867	4	1699.217	18.926	.000
	Within Groups	2244.500	25	89.780		
	Total	9041.367	29			

The ANOVA table from SPSS.

Of all the information presented in the ANOVA table, the major interest of the researcher will most likely be focused on the value located in the "Sig." column, because this is the exact significance level of the ANOVA. If the number (or numbers) found in this column is (are) less than the critical value of alpha (α) set by the experimenter, then the effect is said to be significant. Since this value is usually set at .05, any value less than that will result in significant effects, while any value greater than this value will result in nonsignificant effects. In the example shown in the previous figure, the exact significance is .000, so the effects would be statistically significant. (As discussed earlier in this text, the exact significance level is not really zero, but is some number too small to show up in the number of decimals presented in the SPSS output.)

In this procedure, finding significant effects implies that the means differ more than would be expected by chance alone. In terms of the previous experiment, it would mean that the treatments were not equally effective. This table does not tell the researcher anything about what the effects were, just that there most likely were real effects.

If the effects are found to be nonsignificant, then the differences between the means are not great enough to allow the researcher to rule out a chance or sampling error explanation. In that case, no further interpretation is attempted.

When the effects are significant, the means must then be examined in order to determine the nature of the effects. There are procedures, called post-hoc tests, to assist the researcher in this task, but often the reason is fairly obvious by looking at the size of the various means. For example, in the preceding analysis, Gestalt Therapy and Behavior Therapy were the most effective in terms of mean improvement.

In the case of significant effects, a graphical presentation of the means can sometimes assist in analysis. The following figure shows a graph of mean values from the preceding analysis.

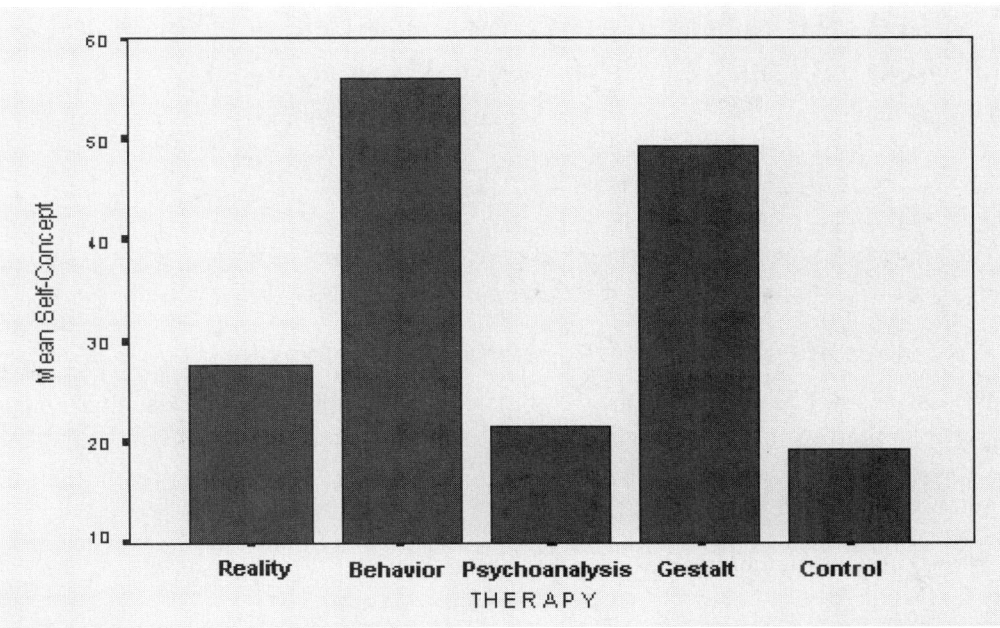

Mean self-concept for the five types of therapy.

21-3 Hypothesis Testing Theory underlying ANOVA

The theory of ANOVA helps explain why the ANOVA hypothesis testing procedure works to simultaneously find effects among any number of means.

First, a review of the sampling distribution is necessary. (If you have difficulty with this summary, please go back and read the Chapter 17, "The Sampling Distribution.")

A sample is a finite number (N) of scores. Sample statistics are numbers that describe the sample. Example statistics are the mean ($\overline{X}$), mode (M_o), median (M_d), and standard deviation (s_X).

Probability models exist in a theoretical world where complete information is unavailable. As such, they can never be known except in the mind of the mathematical statistician. If an infinite number of infinitely precise scores were taken, the resulting distribution would be a probability model of the population. Models of scores are characterized by parameters. Two common parameters are μ and σ.

Sample statistics are used as estimators of the corresponding parameters in the population model. For example, the mean and standard deviation of the sample are used as estimates of the corresponding model parameters μ_X and σ_X.

The sampling distribution is a distribution of a sample statistic. It is a model of a distribution of scores, like the population distribution, except that the scores are not raw scores, but statistics. It is a thought experiment: "What would the world be like if a person repeatedly took samples of size N from the population distribution and computed a particular statistic each time?" The resulting distribution of statistics is called the sampling distribution of that statistic.

The sampling distribution of the mean is a special case of a sampling distribution. It is a distribution of sample means, described with the parameters $\mu_{\overline{x}}$ and $\sigma_{\overline{x}}$. These parameters are closely related to the parameters of the population distribution, the relationship being described by the Central Limit Theorem. This theorem essentially states that the mean of the sampling distribution of the mean ($\mu_{\overline{x}}$) equals the mean of the model of scores (μ_X), and that the standard error of the mean ($\sigma_{\bar{x}}$) equals the theoretical standard deviation of the model of scores (σ_X.) divided by the square root of N. These relationships may be summarized as:

$$\mu_{\bar{X}} = \mu_X$$

$$\sigma_{\bar{X}} = \frac{\sigma_X}{\sqrt{N}}$$

Critical relationships between the parameters of the sampling distribution of the mean and the model of scores.

21-3a Two Ways of Estimating the Population Parameter σ_X^2

When the data have been collected from more than one sample, there exist two independent methods of estimating the parameter σ_X^2—the between method and the within method. The collected data are usually first described with sample statistics, as demonstrated in the following example:

Report

Self-Concept

Reality	Mean	27.50
	Variance	111.500
Behavior	Mean	55.83
	Variance	194.967
Psychoanalysis	Mean	21.33
	Variance	54.667
Gestalt	Mean	49.17
	Variance	64.167
Control	Mean	19.00
	Variance	23.600
Total	Mean	34.57
	Variance	311.771

Means and variances of self-concept for the five levels of therapy.

The total mean and variance is the mean and variance of all 30 scores in the sample.

The Within Method

Since each of the sample variances may be considered an independent estimate of the parameter σ_X^2, finding the mean of the variances provides a method of combining the separate estimates of σ_X^2 into a single value. The resulting statistic is called the **Mean Squares Within,** often represented by **MS_W**. It is called the within method because it computes the estimate by combining the variances within each sample. In the previous example, the Mean Squares Within would be equal to 89.78 or the mean of 111.5, 194.97, 54.67, 64.17, and 23.6. The following formula defines the Mean Squares Within as the mean of the variances:

$$MS_W = \bar{s}^2$$

Computing the Mean Squares Within.

The Between Method

The parameter σ_X^2 may also be estimated by comparing the means of the different samples, but the logic is slightly less straightforward and employs both the concept of the sampling distribution and the Central Limit Theorem.

First, the standard error of the mean squared ($\sigma_{\bar{x}}^2$) is the theoretical variance of a distribution of sample means. In a real-life situation where there is more than one sample, the variance of the sample means may be used as an estimate of the standard error of the mean squared ($\sigma_{\bar{x}}^2$). This is analogous to the situation where the variance of the sample (s^2) is used as an estimate of σ^2.

In this case, the sampling distribution consists of an infinite number of means and the real-life data consists of A (in this case 5) means. The computed statistic is thus an estimate of the theoretical parameter.

The relationship between the standard error of the mean and the sigma of the model of scores expressed in the Central Limit Theorem may now be used to obtain an estimate of σ^2. First, both sides of the equation are squared and then multiplied by N, resulting in the following transformation:

$$\sigma_{\bar{x}} = \frac{\sigma_x}{\sqrt{N}}$$

$$\sigma_{\bar{x}}^2 = \frac{\sigma_x^2}{N}$$

$$N * \sigma_{\bar{x}}^2 = \sigma_x^2$$

Transforming the relationship between the standard error of the mean and the theoretical standard deviation of scores.

Thus the variance of the population may be found by multiplying the standard error of the mean squared ($\sigma_{\bar{x}}^2$) by N, the size of each sample.

Since the variance of the means, $s_{\bar{x}}^2$, is an estimate of the standard error of the mean squared, $\sigma_{\bar{x}}^2$, the variance of the population, σ_X^2, may be estimated by multiplying the size of each sample, N, by the variance of the sample means. This value is called the **Mean Squares Between** and is often symbolized by **MS_B.** The computational procedure for MS_B is presented here:

$$MS_B = N * s_{\bar{x}}^2$$

Formula to compute Mean Squares Between.

The expressed value is called the Mean Squares Between because it uses the variance *between* the sample means to compute the estimate. Using this procedure on the example data yields:

$$MS_B = N * s_{\bar{x}}^2$$
$$MS_B = 6 * 283.21$$
$$MS_B = 1699.28$$

Example computation of Mean Squares Between.

At this point it has been established that there are two methods of estimating σ^2, Mean Squares Within and Mean Squares Between. It could also be demonstrated that these estimates are independent. Because of this independence, when both mean squares are computed using the same data set, different estimates will result. For example, in the presented data $MS_W = 89.78$ while $MS_B = 1699.28$. This difference provides the theoretical background for the F-ratio and ANOVA.

21-3b The F-Ratio

A new statistic, called the **F-ratio,** is computed by dividing the MS_B by MS_W.

$$F = \frac{MS_B}{MS_W}$$

Formula for F observed.

Using the example data described earlier, the computed F-ratio becomes

$$F = \frac{MS_B}{MS_W}$$
$$F = \frac{1699.28}{89.78} = 18.927$$

Applying the formula for F observed.

The F-ratio can be thought of as a measure of how different the means are relative to the variability within each sample. As such, the F-ratio is a measure of the size of the effects. The larger this value, the greater the likelihood that the differences between the means are due to something other than chance alone, namely real effects. How big this F-ratio needs to be in order to make a decision about the reality of effects is the next topic of discussion.

If the difference between the means is due only to chance, that is, there are no real effects, then the expected value of the F-ratio would be one (1.00). This is true because both the numerator and the denominator of the F-ratio are estimates of the same parameter, σ^2. Seldom will the F-ratio be exactly equal to 1.00, however, because the numerator and the denominator are estimates rather than exact values. Therefore, when there are no effects the F-ratio will sometimes be greater or less than one.

To review, the basic procedure used in hypothesis testing is that a model is created in which the experiment is repeated an infinite number of times when there are no effects. A sampling distribution of a statistic is used as the model of what the world would look like if there were no effects. The result of the experiment, also called a statistic, is compared with what would be expected given the model of no effects is true. If the computed statistic is unlikely given the model, then the model is rejected, along with the hypothesis that there were no effects.

In an ANOVA, the F-ratio is the statistic used to test the hypothesis that the effects are real: in other words, that the means are significantly different from one another. Before the details of the hypothesis test may be presented, the sampling distribution of the F-ratio must be discussed.

21-3c The F-Distribution

If the experiment were repeated an infinite number of times, each time computing the F-ratio, and there were no effects, the resulting distribution could be described by the F-distribution. The

F-distribution is a theoretical probability distribution characterized by two parameters, df_1 and df_2, both of which affect the shape of the distribution. Since the F-ratio must always be positive, the F-distribution is asymmetrical, skewed in the positive direction.

The F-ratio, which cuts off various proportions of the distributions, may be found for different values of df_1 and df_2. These F-ratios are called F_{crit} values and may be found with the Probability Calculator by selecting F-Distribution; entering the appropriate values for degrees of freedom and probabilities; and then clicking the arrow pointing to the right.

Following are two examples of using the Probability Calculator to find an F_{crit}. In the first, $df_1 = 10$, $df_2 = 25$, and alpha = .05; and in the second, with $df_1 = 1$, $df_2 = 5$, and alpha = .01. In the first example, the value of $F_{crit} = 2.437$, and in the second, $F_{crit} = 16.258$.

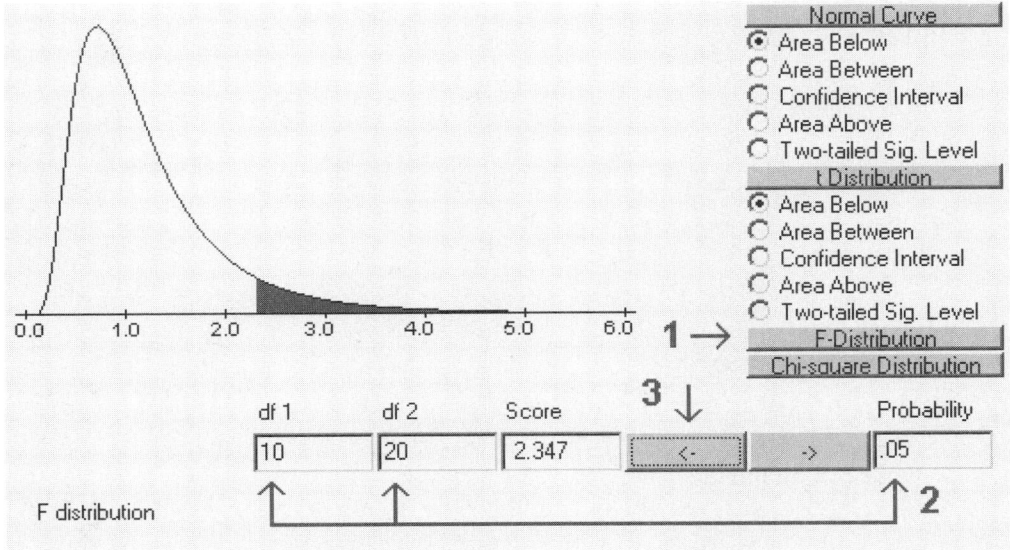

The .05 critical value of an F-distribution with 10 and 20 degrees of freedom.

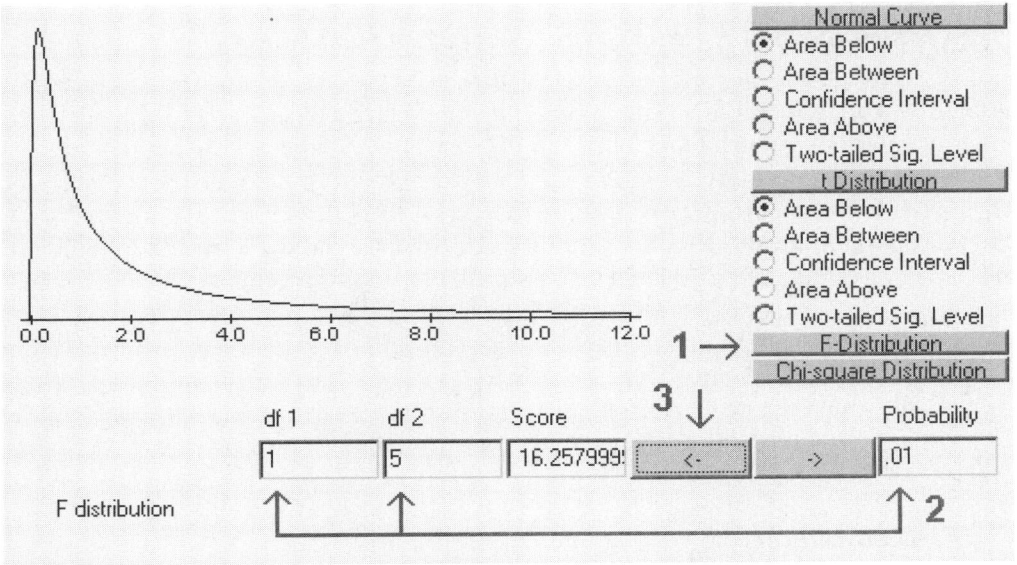

The .01 critical value of an F-distribution with 1 and 5 degrees of freedom.

21-3d Finding Exact Significance Levels for the F-Ratio

The exact significance level of any F-ratio relative to a given F-distribution can be found using the Probability Calculator. Select the F-Distribution; enter the appropriate degrees of freedom and the F-ratio that was found; and then click the arrow pointing to the right. The use of the Probability Calculator to find the exact significance level for the example F-ratio (18.962) described earlier in this chapter is presented here:

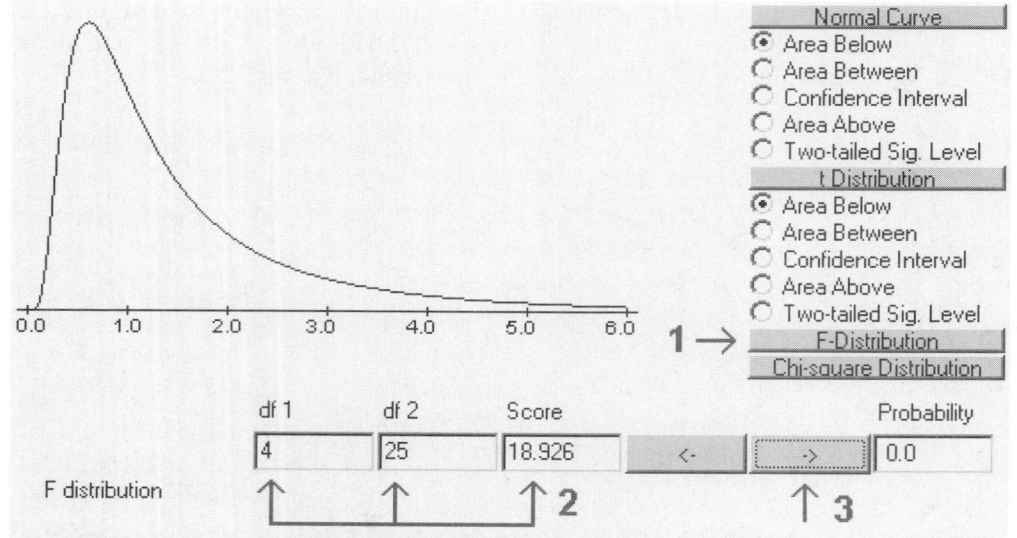

The .01 and .05 critical values of an F-distribution with 4 and 25 degrees of freedom.

As in the SPSS ANOVA output of this data you saw earlier, the probability or exact significance level is too small to register within the decimals allowed in the display, so a value of zero is presented. The null hypothesis would be rejected and the alternative hypothesis accepted, because the exact significance level is less than alpha. The exact significance level found using the Probability Calculator and SPSS should be similar.

21-3e The Distribution of F-Ratios When There Are Real Effects

Theoretically, when there are no real effects, the F-distribution is an accurate model of the distribution of *F-ratios*. The F-distribution will have the parameters $df_1 = A - 1$ (where $A - 1$ is the number of different groups minus one) and $df_2 = A(N - 1)$, where A is the number of groups and N is the number in each group. In this case, an assumption is made that sample size is equal for each group. For example, if five groups of six subjects each were run in an experiment, and there were no effects, the F-ratios would be distributed with $df_1 = A - 1 = 5 - 1 = 4$ and $df_2 = A(N - 1) = 5(6 - 1) = 5 * 5 = 25$. You can see a visual representation of this in the following figure:

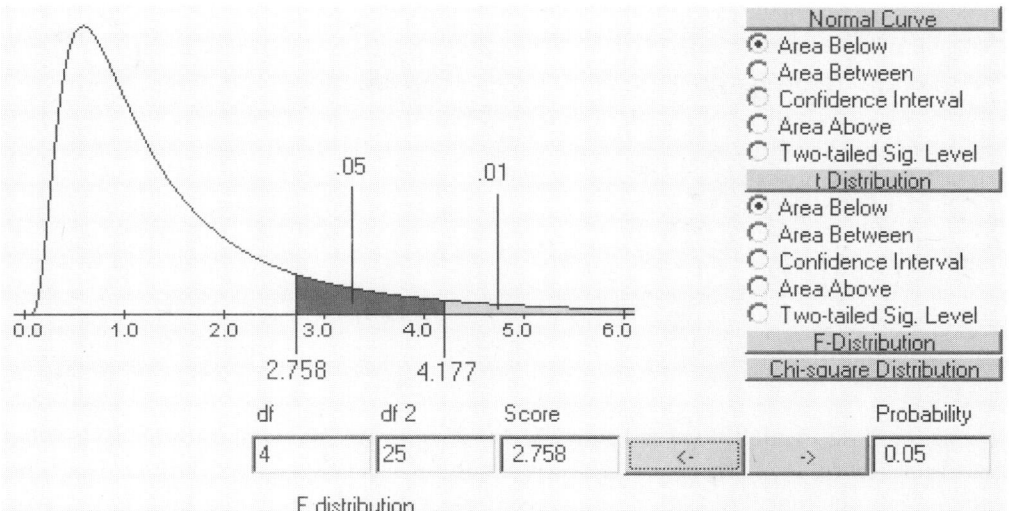

The .01 and .05 critical values of an F-distribution with 4 and 25 degrees of freedom.

When there are real effects, that is, the means of the groups are different due to something other than chance, the F-distribution no longer describes the distribution of F-ratios. In almost all cases, the observed F-ratio will be larger than would be expected when there were no effects. Take a look at the rationale for this situation.

First, an assumption is made that any effects are an additive transformation of the score. That is, the scores for each subject in each group can be modeled as a constant (a_a – the effect) plus error (e_{ae}). A score would appear as

$$X_{ae} = a_a + e_{ae}$$

where X_{ae} is the score for Subject e in group a, a_a is the size of the effect, and e_{ae} is the size of the error. The e_{ae}, or error, is different for each subject, while a_a is constant within a given group.

As described in the chapter on transformations, an additive transformation changes the mean but not the standard deviation or the variance. Because the variance of each group is not changed by the nature of the effects, the Mean Squares Within, as the mean of the variances, is not affected. The Mean Squares Between, as N times the variance of the means, will in most cases become larger because the variance of the means will most likely increase.

Imagine three individuals taking a test. An instructor first finds the variance of the three scores. He then adds five points to one random individual score and subtracts five from another random individual score. In most cases the variance of the three test score will increase, although it is possible that the variance could decrease if the points were added to the individual with the lowest score and subtracted from the individual with the highest score. If the constant added and subtracted were 30 rather than 5, then the variance would almost certainly be increased. Thus, the greater the size of the constant, the greater the likelihood of a larger increase in the variance.

With respect to the sampling distribution, the model differs depending upon whether or not there are effects. The difference is presented in the following figure:

```
         No effects                      Real Effects
  Group  Mean  Variance          Group      Mean      Variance
    1     μ      σ²                 1       μ + a₁        σ²
    2     μ      σ²                 2       μ + a₂        σ²
    3     μ      σ²                 3       μ + a₃        σ²
    4     μ      σ²                 4       μ + a₄        σ²
    5     μ      σ²                 5       μ + a₅        σ²
  Mean    μ      σ²                              μ         σ²
  Variance σ²/N                              >σ²/N
```

The variance of the means generally increases when the effects are different from zero.

Since the MS_B usually increases and MS_W remains the same, the F-ratio ($F = MS_B / MS_W$) will most likely increase. If there are real effects, the F-ratio obtained from the experiment will most likely be larger than the critical level from the F-distribution. The greater the size of the effects, the larger the obtained F-ratio is likely to become.

Thus, when there are no effects, the obtained F-ratio will be distributed as an F-distribution that may be specified. If effects exist, the obtained F-ratio will most likely become larger. By comparing the obtained F-ratio with that predicted by the model of no effects, a hypothesis test may be performed to decide on the reality of effects. If the exact significance level of the F-ratio is less than the value set for alpha, the decision will be that the effects are real. If not, then no decision about the reality of effects can be made.

21-4 Similarity of ANOVA and T-Test

When the number of groups (A) equals two (2), application of either the ANOVA or t-test procedures will result in identical exact significance levels. This equality is demonstrated in the example that follows.

Here is the example data for two groups:

Group										$\overline{X}$	S_x^2
1	12	23	14	21	19	23	26	11	16	18.33	28.50
2	10	17	20	14	23	11	14	15	19	15.89	18.11

Computing the T-Test

$$S_{\overline{x}_1} - S_{\overline{x}_2} = \sqrt{\frac{s_1^2 + s_2^2}{N}} = \sqrt{\frac{28.50 + 18.11}{9}} = 2.28$$

$$\overline{X}_1 - \overline{X}_2 = 18.33 - 15.89 = 2.44$$

Performing a t test on the example data.

Finding an exact significance level using the Probability Calculator's Two-tailed Sig Level under the t Distribution with 16 degrees of freedom, a mu equal to 0, sigma equal to 2.28, and the value equal to 2.44 yields a probability or exact significance level of .30.

Computing the ANOVA

$$MS_{Bet} = N\, s_{\bar{x}}^2 = 9 * 2.9768 = 26.79$$

$$MS_{With} = \frac{s_1^2 + s_2^2}{2} = \frac{28.50 + 18.11}{2} = 23.305$$

$$F = \frac{MS_{Bet}}{MS_{With}} = \frac{26.79}{23.305} = 1.15$$

Computing an ANOVA for the example data.

Using the F-Distribution option of the Probability Calculator with values of 1 and 16 for the degrees of freedom and 1.15 for the value results in an exact probability value of .30.

Because the t-test is a special case of the ANOVA and will always yield similar results, most researchers perform the ANOVA because the technique is much more powerful in analysis of complex experimental designs.

21-5 Example of a Nonsignificant One-Way ANOVA

Given the following data for five groups, perform an ANOVA:

Descriptives

	N	Mean	Std. Deviation	Std. Error	95% Confidence Interval for Mean Lower Bound	95% Confidence Interval for Mean Upper Bound
Reality	6	32.65	15.71	6.41	16.16	49.13
Behavior	6	47.71	18.52	7.56	28.27	67.15
Psychoanalysis	6	42.70	26.44	10.79	14.96	70.45
Gestalt	6	55.68	16.84	6.87	38.01	73.35
Control	6	51.77	16.51	6.74	34.45	69.09
Total	30	46.10	19.60	3.58	38.78	53.42

Descriptive statistics from the ANOVA procedure.

The ANOVA summary table that results should look like this:

ANOVA

		Sum of Squares	df	Mean Square	F	Sig.
Self-Concept	Between Groups	1914.087	4	478.522	1.297	.298
	Within Groups	9223.687	25	368.947		
	Total	11137.8	29			

ANOVA summary table.

Since the exact significance level (.298) provided in SPSS output is greater than alpha (.05), the results are not statistically significant.

21-6 Example of a Significant One-Way ANOVA

Given the following data for five groups, perform an ANOVA. Note that the numbers are similar to the previous example except that three has been added to each score in Group 1, six to Group 2, nine to Group 3, twelve to Group 4, and fifteen to Group 5. This is equivalent to adding effects (a_a) to the scores. Note that the means change, but the variances do not.

Descriptives

THERAPY	N	Mean	Std. Deviation	Std. Error	95% Confidence Interval for Mean Lower Bound	95% Confidence Interval for Mean Upper Bound
Reality	6	35.6452	15.7098	6.4135	19.1591	52.1314
Behavior	6	53.7078	18.5234	7.5622	34.2689	73.1466
Psychoanalysis	6	51.7050	26.4358	10.7924	23.9627	79.4472
Gestalt	6	67.6775	16.8385	6.8743	50.0069	85.3481
Control	6	66.7696	16.5057	6.7384	49.4482	84.0910
Total	30	55.1010	21.4489	3.9160	47.0919	63.1102

Table of descriptive statistics.

The SPSS ANOVA output table should look like this:

ANOVA

		Sum of Squares	df	Mean Square	F	Sig.
SELFCON1	Between Groups	4117.950	4	1029.487	2.790	.048
	Within Groups	9223.687	25	368.947		
	Total	13341.6	29			

ANOVA summary table.

In this case, the Sig. value (.048) is less than .05 and the null hypothesis must be rejected. If the alpha level had been set at .01, or even .045, the results of the hypothesis test would not be statistically significant. In classical hypothesis testing, however, there is no such thing as "close"; the results are either significant or not significant. In practice, however, researchers will often report the exact significance level and let the reader set his or her own significance level. When this is done, the distinction between Bayesian and Classical hypothesis testing approaches becomes somewhat blurred. (Personally, I think that anything that gives the reader more information about your data without a great deal of cognitive effort is valuable and should be done. The reader should be aware that many other statisticians oppose the reporting of exact significance levels.)

Summary

Analysis of Variance (ANOVA) is a hypothesis testing procedure that tests whether two or more means are significantly different from each other. A statistic, F, is calculated that measures the size of the effects by comparing a ratio of the differences between the means of the groups to the variability within groups. The larger the value of F, the more likely that there are real effects. The obtained F-ratio is compared to a model of F-ratios that would be found given that there were no effects. If the obtained F-ratio is unlikely given the model of no effects, the hypothesis of no effects is rejected and the hypothesis of real effects is accepted. If the model of no effects could explain the results, then the null hypothesis of no effects must be retained. The exact significance level is the probability of finding an F-ratio equal to or larger than the one found in the study given that there were no effects. If the exact significance level is less than alpha, then you decide that the effects are real; otherwise you decide that chance could explain the results.

When there are only two groups, a two-tailed t-test and ANOVA will produce a similar exact significance level and make similar decisions about the reality of effects. Since ANOVA is a more general hypothesis testing procedure, it is preferred over a t-test.

Chapter

22

Chi-Square and Tests of Contingency Tables

Key Terms

chi-squared distribution
chi-square statistic

Hypothesis tests may be performed on contingency tables in order to decide whether or not effects are present. Effects in a contingency table are defined as relationships between the row and column variables; that is, are the levels of the row variable differentially distributed over levels of the column variables? Significance in this hypothesis test means that interpretation of the cell frequencies is warranted. Nonsignificance means that any differences in cell frequencies could be explained by chance.

Hypothesis tests on contingency tables are based on a statistic called chi-square. Before we get into a discussion of chi-square, let's review contingency tables.

22-1 Review of Contingency Tables

Frequency tables of two variables presented simultaneously are called contingency tables. Contingency tables are constructed by listing all the levels of one variable as rows in a table and the levels of the other variables as columns, and then finding the joint or cell frequency for each cell. The cell frequencies are then summed across both rows and columns. The sums are placed in the margins, the values of which are called marginal frequencies. The lower right-hand corner value contains the sum of either the row or column marginal frequencies, which both must be equal to N.

For example, suppose that a researcher studied the relationship between being HIV positive and the sexual preference of individuals. The study resulted in the following data for thirty male subjects:

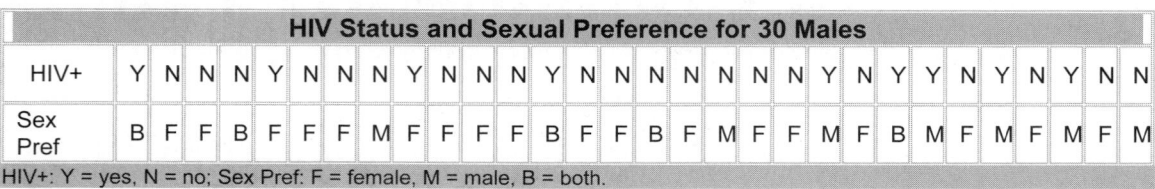

HIV Status and Sexual Preference for 30 Males																													
HIV+	Y	N	N	N	Y	N	N	N	Y	N	N	N	Y	N	N	N	N	N	N	N	Y	N	Y	Y	N	Y	N	Y	N	N
Sex Pref	B	F	F	B	F	F	F	M	F	F	F	F	B	F	F	B	F	M	F	F	M	F	B	M	F	M	F	M	F	M

HIV+: Y = yes, N = no; Sex Pref: F = female, M = male, B = both.

The SPSS data file—coding 1 = Yes and 2 = No for HIV, and 1 = males, 2 = females, and 3 = both for sexual preference—would appear as follows (partial view):

	hiv	sex
16	0	3
17	0	2
18	0	1
19	0	2
20	0	2
21	1	1
22	0	2
23	1	3
24	1	1
25	0	2
26	1	1
27	0	2
28	1	1
29	0	2
30	0	1

Entering the data in SPSS to compute a contingency table.

A contingency table and chi-square hypothesis test of independence can be generated in SPSS by selecting Analyze/Descriptive Statistics/Crosstabs as the following figure shows:

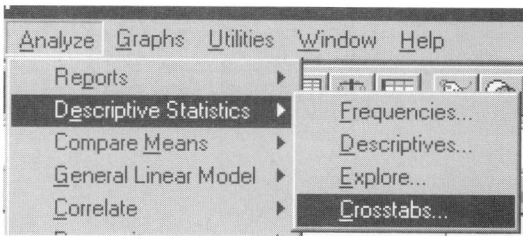

Finding the SPSS Crosstabs command.

Then select the options indicated in the following figure.

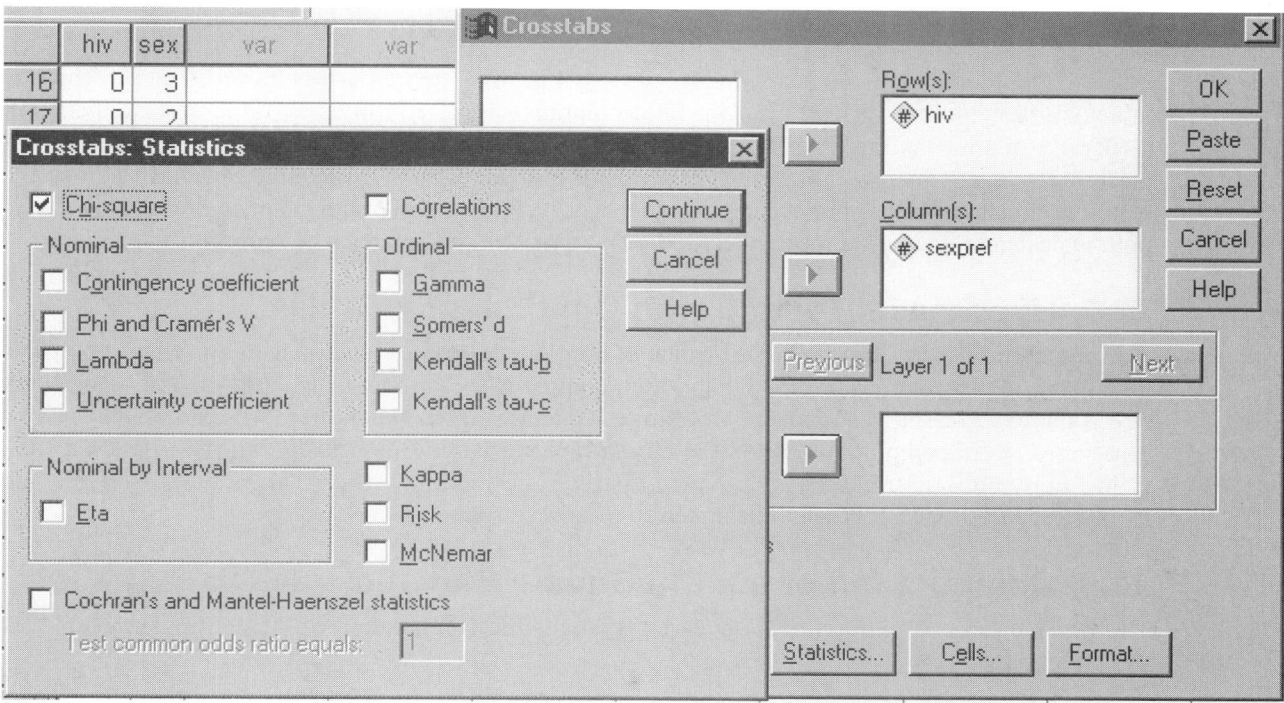

Selecting the variables and optional chi-square statistics in SPSS Crosstabs.

The resulting output tables should look like these:

HIV * SEXPREF Crosstabulation

Count

		SEXPREF			
		Males	Females	Both	Total
HIV	No	3	16	2	21
	Yes	4	2	3	9
Total		7	18	5	30

The contingency table produced by the SPSS Crosstabs command.

Chi-Square Tests

	Value	df	Asymp. Sig. (2-tailed)
Pearson Chi-Square	7.657[a]	2	.022
Likelihood Ratio	7.803	2	.020
Linear-by-Linear Association	.062	1	.803
N of Valid Cases	30		

a. 4 cells (66.7%) have expected count less than 5. The minimum expected count is 1.50.

Chi-square test produced by the SPSS Crosstabs command.

The Pearson chi-square value in the Asymp. Sig. (two-tailed) column is 0.022, which is less than .05, indicating that the rows and columns of the contingency table are independent. Generally this means that it is worthwhile to interpret the cells in the contingency table. In this particular case, it means that being HIV positive or not is not distributed similarly across the different levels of sexual preference. In other words, males who prefer other males or who prefer both males and females are more likely to be HIV positive than males who prefer only females.

22-2 Hypothesis Testing with Contingency Tables

The procedure used to test the significance of contingency tables is similar to all other hypothesis tests. That is, a statistic is computed and then compared to a model of what the world would look like if the experiment was repeated an infinite number of times when there were no effects. In this case the statistic computed is called the **chi-square statistic.** This statistic will be discussed first, followed by a discussion of its theoretical distribution. Finding critical values of chi-squared and its interpretation will conclude the chapter.

22-3 Computation of the Chi-Square Statistic

The first step in computing the chi-square statistic is the computation of the contingency table, as shown here:

	SEXPREF			
	Males	Females	Both	
HIV+	4	2	3	9
Not HIV+	3	16	2	21
	7	18	5	30

A contingency table with absolute frequencies.

The next step in computing the chi-square statistic is the computation of the expected cell frequency for each cell. To do this, multiply the marginal frequencies for the row and column (row and column totals) of the desired cell and then divide by the total number of observations. The formula for computation can be represented as follows:

Expected Cell Frequency = (Row Total * Column Total) / N

For example, computation of the expected cell frequency for HIV+ Males would proceed as follows:

Expected Cell Frequency = (Row Total * Column Total) / N

= (9 * 7) / 30 = 2.1

You can see the cell we are working with in the following table:

| | SEXPREF | | | |
	Males	Females	Both	Row Total
HIV+	Expected (9*7)/30 = 2.1			9
Not HIV+				
Column Total	7			N = 30

Calculating the expected cell frequency.

Using the same procedure to compute all the expected cell frequencies results in the following table:

| | SEXPREF | | | |
	Males	Females	Both	
Observed	4	2	3	9
Expected	2.1	5.4	1.5	
Observed	3	16	2	21
Expected	4.9	12.6	3.5	
	7	18	5	30

Expected cell frequency for all cells.

Note that the sum of the expected row totals is the same as the sum of the observed row totals; the same holds true for the column totals.

The next step is to subtract the expected cell frequency from the observed cell frequency for each cell. This value gives the amount of deviation or error for each cell. Adding these to the preceding table results in the following:

	SEXPREF			
	Males	Females	Both	
Observed	4	2	3	9
Expected	2.1	5.4	1.5	
O-E	1.9	-3.4	1.5	
Observed	3	16	2	21
Expected	4.9	12.6	3.5	
O-E	-1.9	3.4	- 1.5	
	7	18	5	30

Observed, expected, and observed minus expected cell frequency.

Note also that the sum of the observed minus expected for both the rows and columns equals zero.

Following this, the difference computed in the last step is squared, resulting in the following table:

	SEXPREF			
	Males	Females	Both	
Observed	4	2	3	9
Expected	2.1	5.4	1.5	9
O-E	1.9	-3.4	1.5	0
$(O-E)^2$	3.61	11.56	2.25	
Observed	3	16	2	21
Expected	4.9	12.6	3.5	21
O-E	-1.9	3.4	-1.5	0
$(O-E)^2$	3.61	11.56	2.25	
	7	18	5	30

The observed, expected, observed minus expected, and observed minus expected squared cell frequency.

Each of the squared differences is then divided by the expected cell frequency for each cell, resulting in the following table:

	SEXPREF			
	Males	Females	Both	
Observed	4	2	3	9
Expected	2.1	5.4	1.5	9
O-E	1.9	-3.4	1.5	0
$(O-E)^2$	3.61	11.56	2.25	
$(O-E)^2/E$	1.72	2.14	1.5	
Observed	3	16	2	21
Expected	4.9	12.6	3.5	21
O-E	-1.9	3.4	-1.5	0
$(O-E)^2$	3.61	11.56	2.25	
$(O-E)^2/E$	.74	.92	.64	
	7	18	5	30

Observed, expected, observed minus expected, observed minus expected squared, and observed minus expected squared divided by expected cell frequency.

The chi-square statistic is computed by summing the last row of each cell in the preceding table, the formula being represented by

$$X^2_{Obs} = \sum_{cells} \frac{(O-E)^2}{E}$$

Computational formula for chi-square.

This computation for the example table would result in the following:

$X^2_{Obs} = 1.72 + 2.14 + 1.50 + .74 + .92 + .64 = 7.66$

Note that this value is within rounding error of the value for chi-square computed by the SPSS earlier in this chapter.

22-4 The Theoretical Distribution of Chi-Square When No Effects Are Present

The distribution of the chi-square statistic may be specified given the preceding experiment were conducted an infinite number of times and the effects were not real. The resulting distribution is called the **chi-squared distribution.** The chi-squared distribution is characterized by a parameter called the degrees of freedom (df) that determines the shape of the distribution. Two chi-squared distributions are presented here, each with a different value for the degrees of freedom parameter.

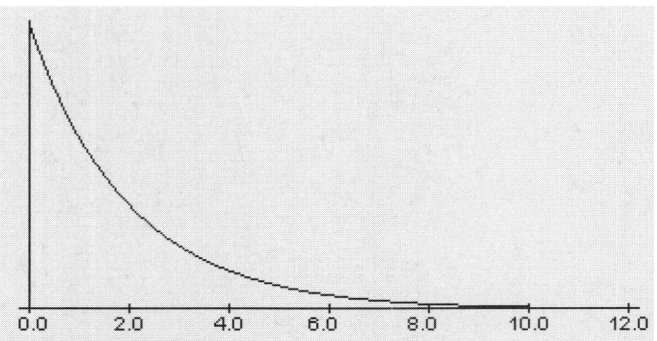

A chi-squared distribution with df = 2.

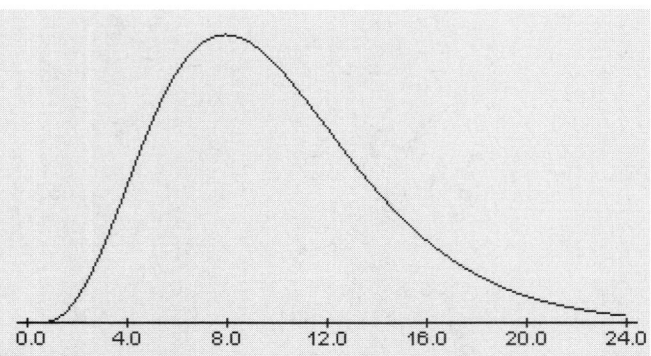

A chi-squared distribution with df = 10.

The degrees of freedom in the example chi-squared distribution is computed by multiplying one minus the number of rows, times one minus the number of columns, or:

df = (#Rows – 1) * (#Columns – 1)

In the example problem, the degrees of freedom is equal to (2 – 1) * (3 – 1) = 1 * 2 = 2.

22-4a Finding the Exact Significance Level for a Chi-Square Statistic

The exact significance level for a chi-square statistic can be found using the Probability Calculator. Select Chi-square Distribution; enter 2 in the df box and 7.66 in the Score box; and then click the right-facing arrow, as the following figure illustrates:

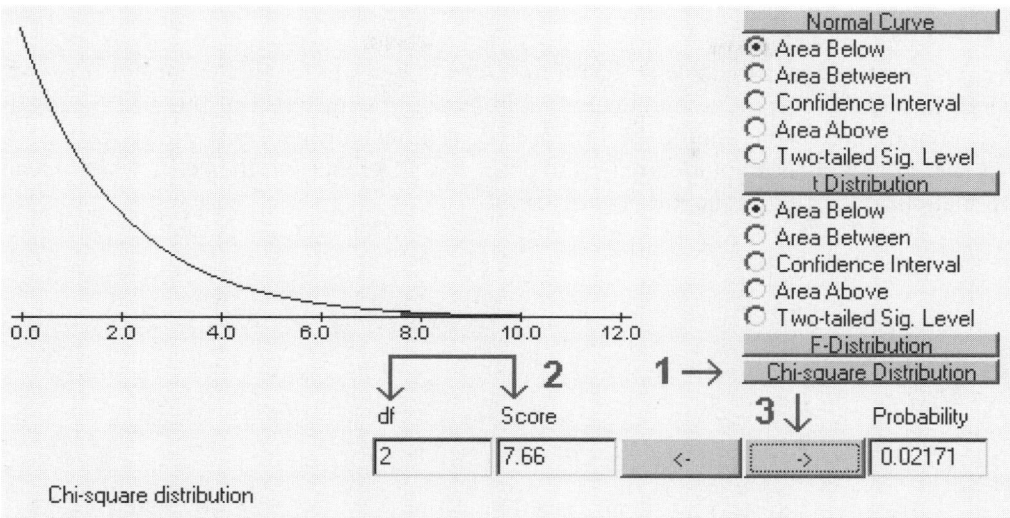

Finding the exact significance level of a chi-square statistic using the Probability Calculator.

The exact significance level computed by the Probability Calculator (.0217) agrees within rounding error of the value computed by SPSS (.022) earlier. In both cases the null hypothesis would be rejected.

22-5 Interpreting a Significant Chi-Square

The interpretation of the cell frequencies may be guided by the amount each cell contributes to the chi-square statistic, as seen in the $(O - E)^2 / E$ value. In general, the larger the difference between the observed and expected values, the greater this value. In the example data, it can be seen that the homosexual males had a greater incidence of being HIV positive (Observed = 4, Expected = 2.1) than would be expected by chance alone, while heterosexual males had a lesser incidence (Observed = 2, Expected = 5.4). This sort of evidence could direct the search for the causes of HIV.

	SEXPREF			
	Males	Females	Both	
Observed	4	2	3	9
Expected	2.1	5.4	1.5	
$(O-E)^2/E$	1.72	2.14	1.5	
Observed	3	16	2	21
Expected	4.9	12.6	3.5	
$(O-E)^2/E$	.74	.92	.64	
	7	18	5	30

A contingency table with cell values necessary to compute the chi-square statistic.

Summary

The chi-squared test of significance is useful as a tool to determine whether or not it is worth the researcher's effort to interpret a contingency table. A significant result of this test means that the cells of a contingency table should be interpreted. A nonsignificant test means that no effects were discovered and chance could explain the observed differences in the cells, which means that an interpretation of the cell frequencies is not useful.

Chapter

23

Testing a Single Correlation Coefficient

The hypothesis tested when testing a single correlation coefficient is that a linear relationship exists between two variables, x and y, as measured by the correlation coefficient (r). The null hypothesis states that no linear relationship exists between the two variables. As in all hypothesis tests, the goal is to reject the null hypothesis and accept the alternative hypothesis; in other words, to decide that an effect, in this case a relationship, exists.

23-1 The Hypothesis and Nature of the Effects

Suppose a study was performed which examined the relationship between life-satisfaction and attitude toward boxing. The attitude toward boxing is measured by the following statement on a questionnaire:

I enjoy watching a good boxing match.

Life-satisfaction is measured with the following statement:

I am pretty much satisfied with my life.

Both items were measured with the following scale:

1 = Strongly Disagree 2 = Disagree 3 = No Opinion 4 = Agree 5 = Strongly Agree

The questionnaire was given to N = 33 people. The obtained correlation coefficient between these two variables was r = -.30. Because the correlation is negative, we conclude that the people who said they enjoyed watching a boxing match were less satisfied with their lives. The corollary, that individuals who said they were satisfied with their lives did not say they enjoyed watching boxing, is also true. On the basis of this evidence the researcher argues that there is a relationship between the two variables.

Before she could decide that there was a relationship, however, a hypothesis test had to be performed to negate, or at least make improbable, the hypothesis that the results were due to chance. The ever-present devil's advocate argues that there really is no relationship between the two variables; the obtained correlation was due to chance. The researcher just happened to select 33 people who had a negative correlation between these two variables. If another sample were taken, the correlation was just as likely to be positive and just as large. Furthermore, if a sample of infinite size (population) was taken and the correlation coefficient computed, the true correlation coefficient would be 0.0. In order to answer this argument, a hypothesis test is needed.

23-2 The Model of No Effects

The model of no effects is described by the sampling distribution of a correlation coefficient. In a thought experiment, the study is repeated an infinite number of times using the same two questions and a different sample of 33 individuals each time, assuming the null hypothesis is true. Computing the correlation coefficient each time results in a sampling distribution of the correlation coefficient. This distribution of correlation coefficients can be graphed in a theoretical relative frequency distribution that is similar to the following:

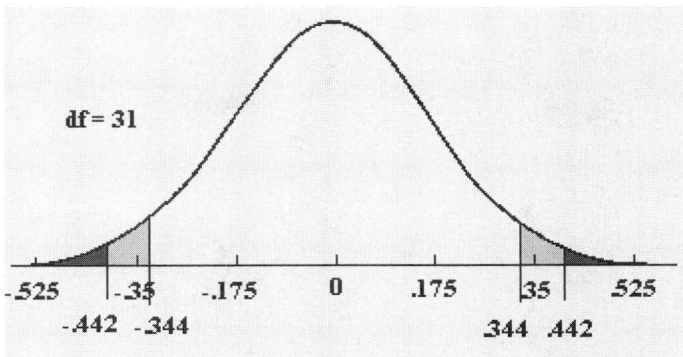

The distribution of correlation coefficients with df = 31.

Note that this distribution looks like a normal distribution. It could not be normal, however, because the scores are limited to the range of -1.0 to 1.0.

Because the sampling distribution of the correlation coefficient has a unique shape, a computer program, which is included in this book, is used to find values that cut off a given proportion of area. *(Note: To access this program, you will need to go to the online version of this text.)* To use the program, you must first find the degrees of freedom using the following formula:

$$df = N - 2$$

Enter the degrees of freedom in the df = box, as shown in the following figure, and, click the Find button.

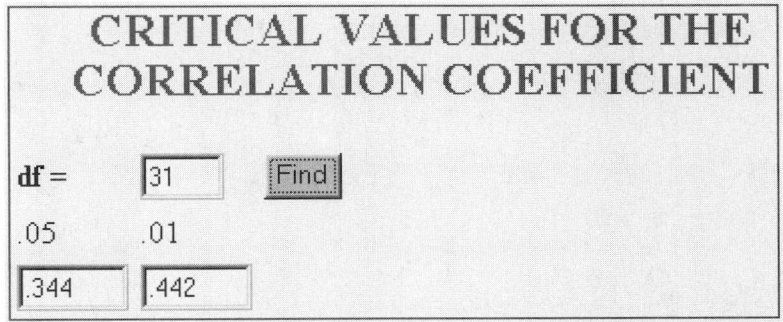

Finding critical values for the correlation coefficient.

These are the same values that appear in the sampling distribution of the correlation coefficient just presented. The values appearing in the row corresponding to the degrees of freedom are areas (probabilities) falling below the tail(s) of the distribution. In the previous example, .95 area falls between correlations of -.344 and .344, and .99 area between -.442 and .442.

23-3 Comparing the Obtained Statistic with the Model

The obtained correlation coefficient is now compared with what would be expected given the model created under the null hypothesis. In the earlier example, the value of -.30 falls inside the critical values of -.344 and .344 that were found using the computer program. Because the obtained correlation coefficient did not fall in the tails of the distribution under the null hypothesis, the model and the corresponding null hypothesis must be retained. The model of no effects could explain the results. The correlation coefficient is not significant at the .05 level.

If the obtained correlation coefficient had been -.55, however, the decision would have been different. In that case the obtained correlation is unlikely, given the model, because it falls in the tails of the sampling distribution. The model and corresponding null hypothesis are rejected as unlikely and the alternative hypothesis, that of a real effect, accepted. The obtained correlation is said to be significant at the .05 level.

Summary

Hypothesis testing using a correlation coefficient to measure the size of the effect tests whether two variables are linearly related to one another. By finding the likelihood of the obtained correlation coefficient given a model of correlation coefficients when there are no effects, a decision can be made about whether the two variables are linearly related.

Appendix

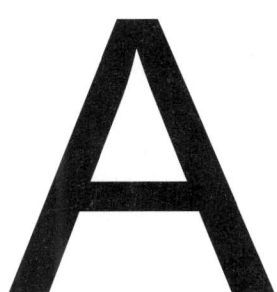

Relational Databases
and Statistical Packages

Key Term

relational database

A relational database is a much more efficient means to store and maintain information that the flat two-dimensional tables of a statistical package. The latest editions of SPSS allow the user to easily construct SPSS data files from many relational databases. This appendix explores relational databases and then shows how to use them to construct SPSS data files.

A-1 Student Attitudes As a Function of Student Attributes, Teacher Attributes, and Environment

An education researcher has been given a grant to study student attitudes toward learning as a function of student attributes, teacher attributes, and teaching environment. Student attributes include such variables as general intelligence, socio-economic status, parental involvement, motivation, absences, gender, age, temperament, and involvement with athletics. Teacher attributes include variables such as age, gender, marital status, educational level, experience, motivation, temperament, and leadership characteristics. Teaching environment includes variables such as class size, lighting, comfort, available technology, and age of building.

The researcher gives an attitude survey consisting of 15 questions that are answered on a scale of 1 = strongly disagree to 5 = strongly agree. The items are to be combined into three scales, one of nine items, one of four items, and one of two items. The researcher wants to construct a prediction model using multivariate statistical methods to predict student attitudes on the three scales from the various student, teacher, and environmental attributes.

The researcher collects data from every student in every class in five schools. If there is an average of 500 students at each school with six classes each, the data file will consist of 5 * 500 * 6 = 15,000 entries or rows in the data file. If there were 25 attribute variables and 15 survey variables, the total number of entries in the data matrix would be 15,000 * (25 + 15) = 600,000. It is going to take a small army of data collectors and data entry operators to create this data file.

The researcher can automate some of the data entry. The survey items can be collected on computer forms that can be read directly into the computer. The students could also enter a code for the teacher and the classroom on the form. It might be difficult or impossible, however, for the student to enter complete information about the teacher, classroom, and personal variables. How many students have an accurate memory of their absences? How many know their general intelligence? This information might exist in other data sources, but the student is not going to be able to enter this on the computer form.

The researcher recognizes that there is a great deal of duplication in the additional information that must be entered in the variables in order to complete the data file. Each classroom, for example, should have the same information entered as environmental variables. The same should happen for each teacher. Because the teachers move around to different classrooms in the schools, the teacher and environmental variables are not completely confounded. Data entry could be further automated using "cut and paste" methods, but these can be tedious, and better methods exist to construct this type of data file.

The end result of the process is a data table that appears as follows:

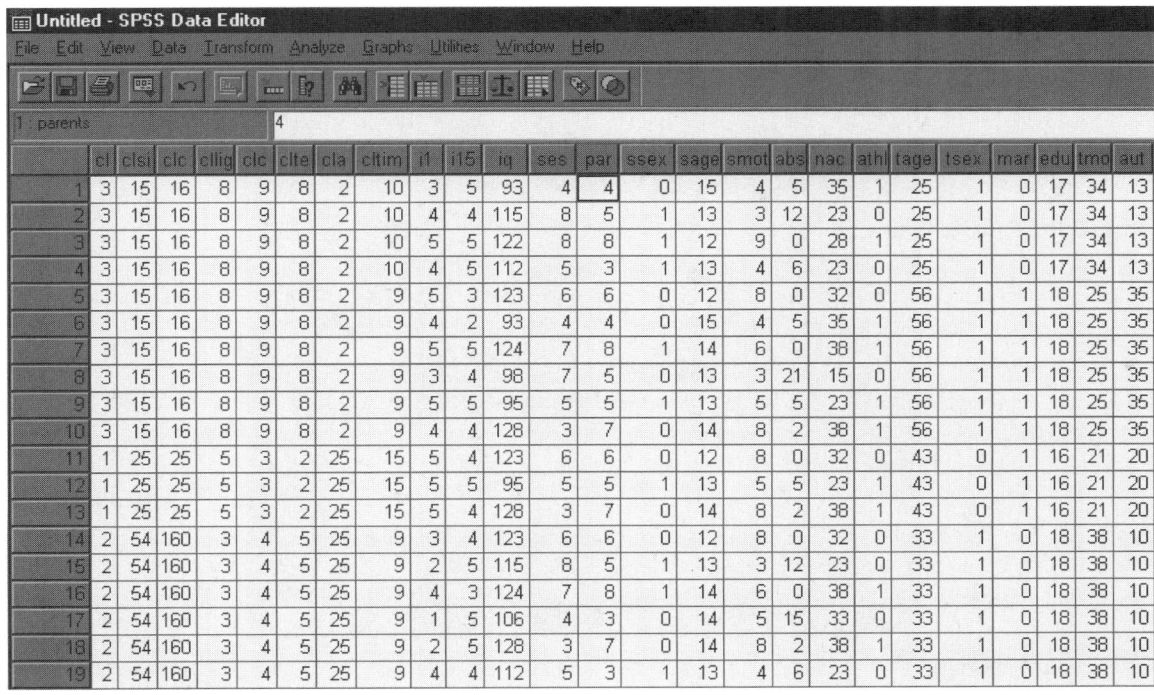

SPSS data file for Student Attitudes

The problem is that the SPSS data file is a flat, two-dimensional, table. Each row is an observation and each column is a variable. This works fine for analyzing data, but it is an inefficient way of storing redundant information.

A-2 A Better Way: Relational Databases

A **relational database** is a series of two-dimensional tables that are linked by an index or key variable. The key allows complex, redundant, two-dimensional tables to be constructed on-the-fly. This system allows the researcher to construct four nonredundant tables, each containing information about one aspect of the study. The tables will be combined into a single table and used as input into the statistical package.

A-2a Setting Up the Database

Any number of different relational database programs would work for the researcher. Microsoft® Access™ has been selected because it is widely available. It is not a "heavy-duty" database in that you might not want to set up a critical e-commerce business using it, but it is more than adequate for our purpose.

Often the relational database that contains the information will have been constructed and maintained by a third party. The job of the researcher is to access and join the tables within the statistical package. Creating a simple relational database in Access is fairly easy and may provide insight into how relational databases function, so before learning how to access a database within a statistical package, you will construct a demonstration database.

A-2b Creating a New Database

Open the Microsoft Access program and when the screen shown in the following figure appears, select Blank Access database and then click OK.

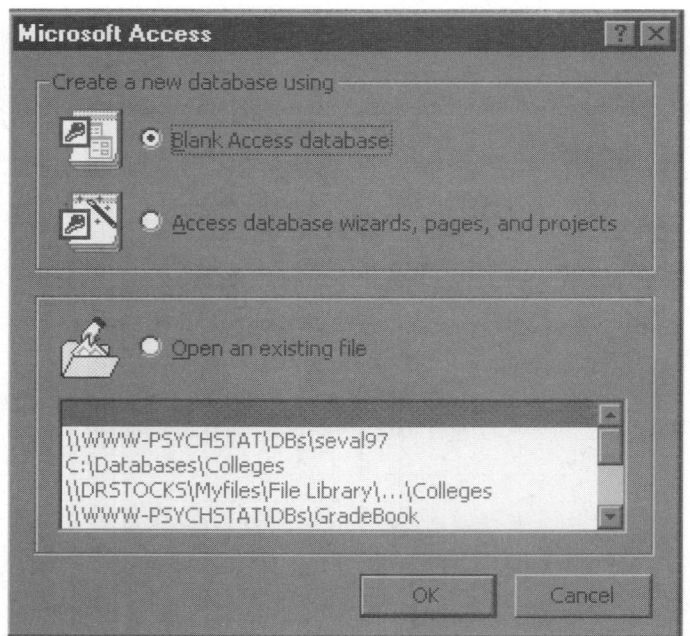

Opening a new database in Access.

On the next screen, select the folder to save the database using the Save in: box. Name the database (in the File name: box) Schools. Then click the Create button to create the database.

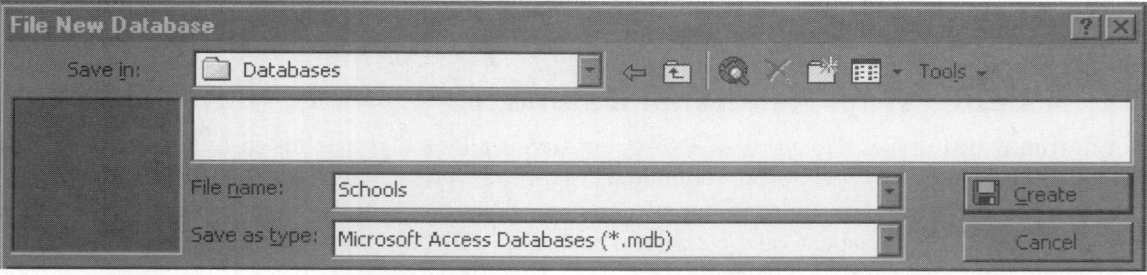

Screen to create and save database in Access.

A-2c Creating the Teacher Table in Access

The database exists now, but contains no information. Four database tables need to be created, one for each specific type of information. The first table will contain information about the teachers. After you clicked Create in the last section, the dialog box shown in the following figure should be open on your screen.

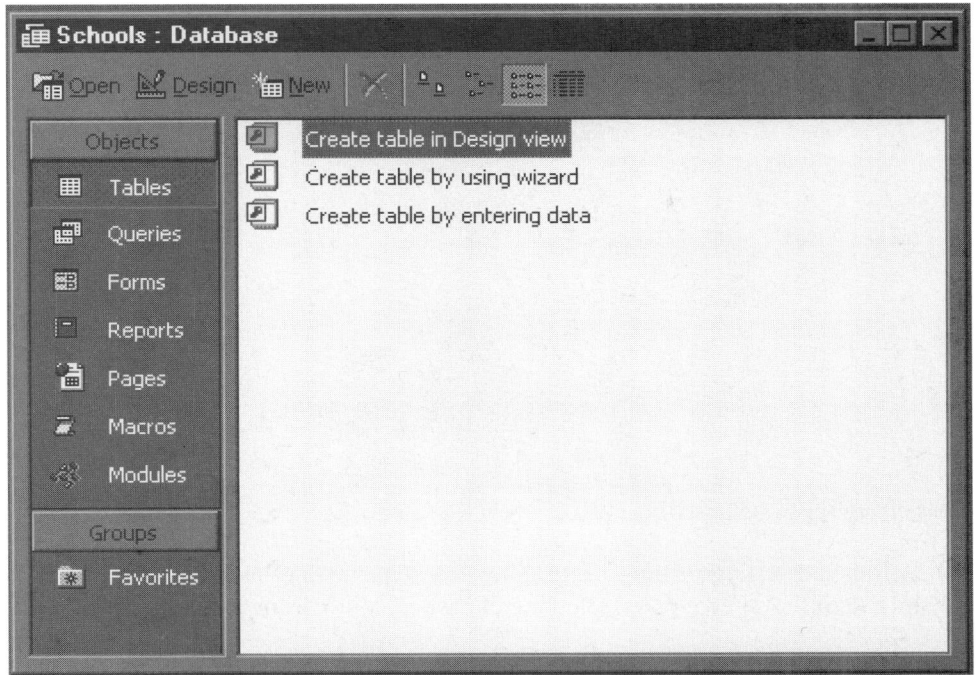

Opening screen in new Access database.

Double-click on Create table in Design View. When the table appears, enter the information shown in the following figure:

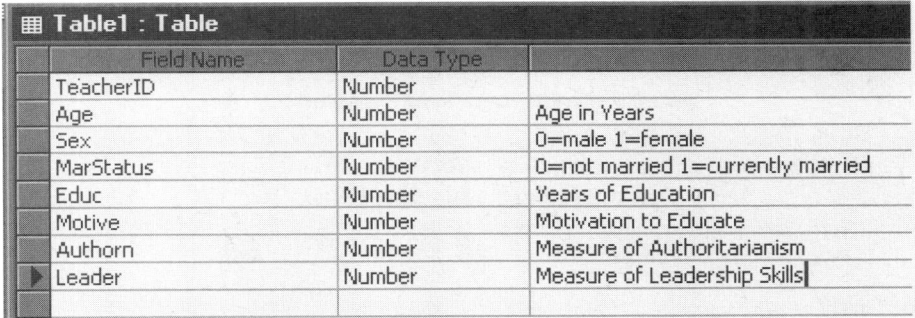

Setting the variables in the Teacher table in design view.

Note: When your cursor is placed in the Data Type column, click the small arrow at the right of the cell to view a dropdown list of possible data types. Select Number for this exercise.

When you have entered all of the Table1 information, right-click the TeacherID field name and select Primary Key from the pop-up menu (shown in the following figure) to make this field the primary key.

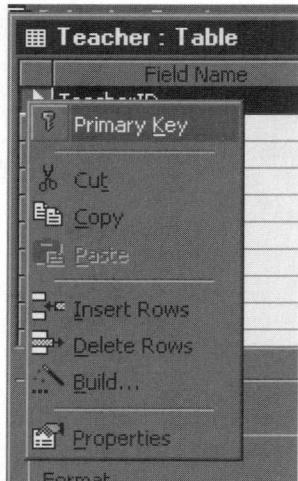

Selecting the primary key in Access.

When finished, change from Design View to Datasheet View by clicking the View icon in the left-hand top of the screen, or by selecting Datasheet View from the View menu.

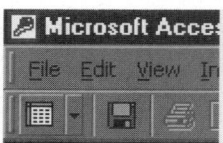

The icon to change views in Access is directly under the File menu.

The computer will ask if you want to save the table before changing views:

Question before changing views in Access.

Click Yes, and a Save As dialog box (shown in the following figure) appears. Type Teacher in the Table Name: box and click OK.

Saving the Teacher table in Access.

Information may be added to the table in the Datasheet View. A new row is created when information is entered in any of the cells. The example table shown in the following figure contains information for five teachers. Enter this information into your Teacher table.

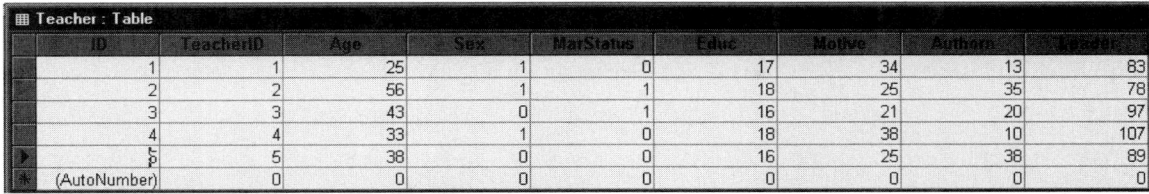

ID	TeacherID	Age	Sex	MarStatus	Educ	Motive	Authorn	Leader
1	1	25	1	0	17	34	13	83
2	2	56	1	1	18	25	35	78
3	3	43	0	1	16	21	20	97
4	4	33	1	0	18	38	10	107
5	5	38	0	0	16	25	38	89
(AutoNumber)	0	0	0	0	0	0	0	0

Datasheet view of Teacher table in Access.

When you've entered all the information to the Teacher table, close the table by clicking the X in the upper right-hand corner of the Teacher:Table screen. The Schools:Database screen reappears, and another table can be created.

A-2d Creating the Other Example Tables

Since our example database requires four tables, the previous procedure must be repeated three more times. First, create the Student table. The Design View is shown in the following figure. Enter the information from it into your new table:

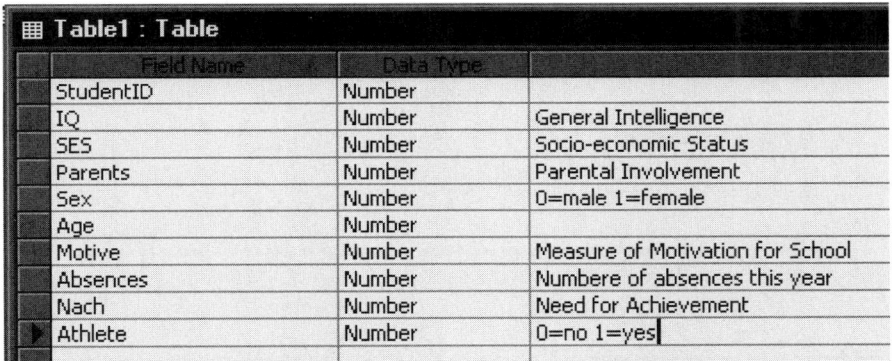

Field Name	Data Type	
StudentID	Number	
IQ	Number	General Intelligence
SES	Number	Socio-economic Status
Parents	Number	Parental Involvement
Sex	Number	0=male 1=female
Age	Number	
Motive	Number	Measure of Motivation for School
Absences	Number	Numbere of absences this year
Nach	Number	Need for Achievement
Athlete	Number	0=no 1=yes

Design view of Student table in Access.

The Student table is constructed with the data shown in the following figure (note that the StudentID would most likely be the student's Social Security number—the examples are not reasonable). Enter the information into your Student table.

	StudentID	IQ	SES	Parents	Sex	Age	Motive	Absences	Nach	Athlete
▶ ⊞	111220001	123	6	6	0	12	8	0	32	0
⊞	111220002	93	4	4	0	15	4	5	35	1
⊞	111220003	115	8	5	1	13	3	12	23	0
⊞	111220004	85	3	6	0	16	2	2	12	0
⊞	111220005	124	7	8	1	14	6	0	38	1
⊞	111220006	98	7	5	0	13	3	21	15	0
⊞	111220007	106	4	3	0	14	5	15	33	0
⊞	111220008	95	5	5	1	13	5	5	23	1
⊞	111220009	128	3	7	0	14	8	2	38	1
⊞	111220010	122	8	8	1	12	9	0	28	1
⊞	111220011	112	5	3	1	13	4	6	23	0
⊞	111220012	108	4	5	0	15	3	18	27	1
✱	0	0	0	0	0	0	0	0	0	0

Datasheet view of Student table in Access.

Next, create the Classroom table. Enter the information shown in the next two figures into your Classroom table.

Table1 : Table

Field Name	Data Type	Description
Classroom	Number	
ClSize	Number	Number of Student in Class
ClCap	Number	Capacity of Classroom
ClLight	Number	Rating of Classroom Lighting 1-10
ClCmft	Number	Rating of Classroom Comfort 1-10
ClTech	Number	Rating of Classroom Technology 1-10
ClAge	Number	Age, in years, of Classroom

Design view of Classroom table in Access.

Classroom : Table

	Classroom	ClSize	ClCap	ClLight	ClCmft	ClTech	ClAge
▶ ⊞	1	25	25	5	3	2	25
⊞	2	54	160	3	4	5	25
⊞	3	15	16	8	9	8	2
⊞	4	54	54	7	7	7	3
✱	0	0	0	0	0	0	0

Datasheet view of Classroom table in Access.

Finally, you need to create the Questionnaire table. Most likely this table would be populated using some kind of automation, such as scantron sheets or by entering information on forms from a Web page. This table is the main table and contains references to the key variable on the other tables. For example, an entry of "2" in the TeacherID variable refers to the teacher identified as "2" in the TeacherID variable in the Teacher table. The variables do not need to have similar names in order to refer to each other, only similar values. An entry called formID is used as the primary key because all other variables are not unique, that is, the same StudentID may (and probably will) appear more than once in this column. The Design view of the Questionnaire table appears in the following figure. Enter this information into your Questionnaire table.

Questionnaire : Table

Field Name	Data Type	
FormID	AutoNumber	
StudentID	Number	Index to StudentID
TeacherID	Number	Index to TeacherID
ClassRoomID	Number	Index to ClassRoomID
ClTime	Number	Class time (24hrs)
I1	Number	First item on attitude questionnaire
I15	Number	Last item on attitude questionnaire

Design view of Questionnaire table in Access.

The Datasheet view of the Questionnaire table is shown in the following figure. Enter the data into your table.

Questionnaire : Table

FormID	StudentID	TeacherID	ClassRoomID	ClTime	I1	I15
1	111220001	2	3	9	5	3
2	111220002	2	3	9	4	2
3	111220005	2	3	9	5	5
4	111220006	2	3	9	3	4
5	111220008	2	3	9	5	5
6	111220009	2	3	9	4	4
7	111220013	2	3	9	5	4
8	111220001	4	2	9	3	4
9	111220003	4	2	9	2	5
10	111220005	4	2	9	4	3
11	111220007	4	2	9	1	5
12	111220009	4	2	9	2	5
13	111220011	4	2	9	4	4
14	111220013	4	2	9	3	5
15	111220002	1	3	10	3	5
16	111220003	1	3	10	4	4
17	111220010	1	3	10	5	5
18	111220011	1	3	10	4	5
19	111220013	1	3	10	3	4
20	111220014	1	3	10	5	2
21	111220001	3	1	15	5	4
22	111220008	3	1	15	5	5
23	111220009	3	1	15	5	4
(AutoNumber)	0	0	0	0	0	0

Datasheet View of the Questionnaire table in Access.

At this point the relational database is complete for the purposes of generating an SPSS data file. The database can either be closed or remain open for the rest of the exercise.

A-3 Reading a Relational Database into an SPSS Data File

Rather than creating or reading an existing SPSS data file, the researcher is going to create a data file by combining the tables in the relational database. To open a relational database as a data file, select File/Open Database/New Query in SPSS, as the following figure shows:

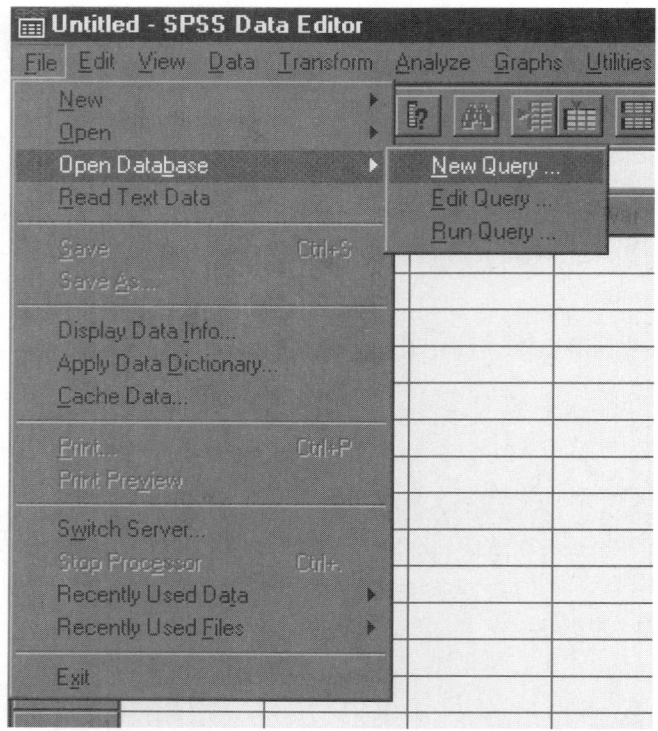

Procedure to open a relational database in SPSS.

The Database Wizard screen, shown in the following figure, should appear:

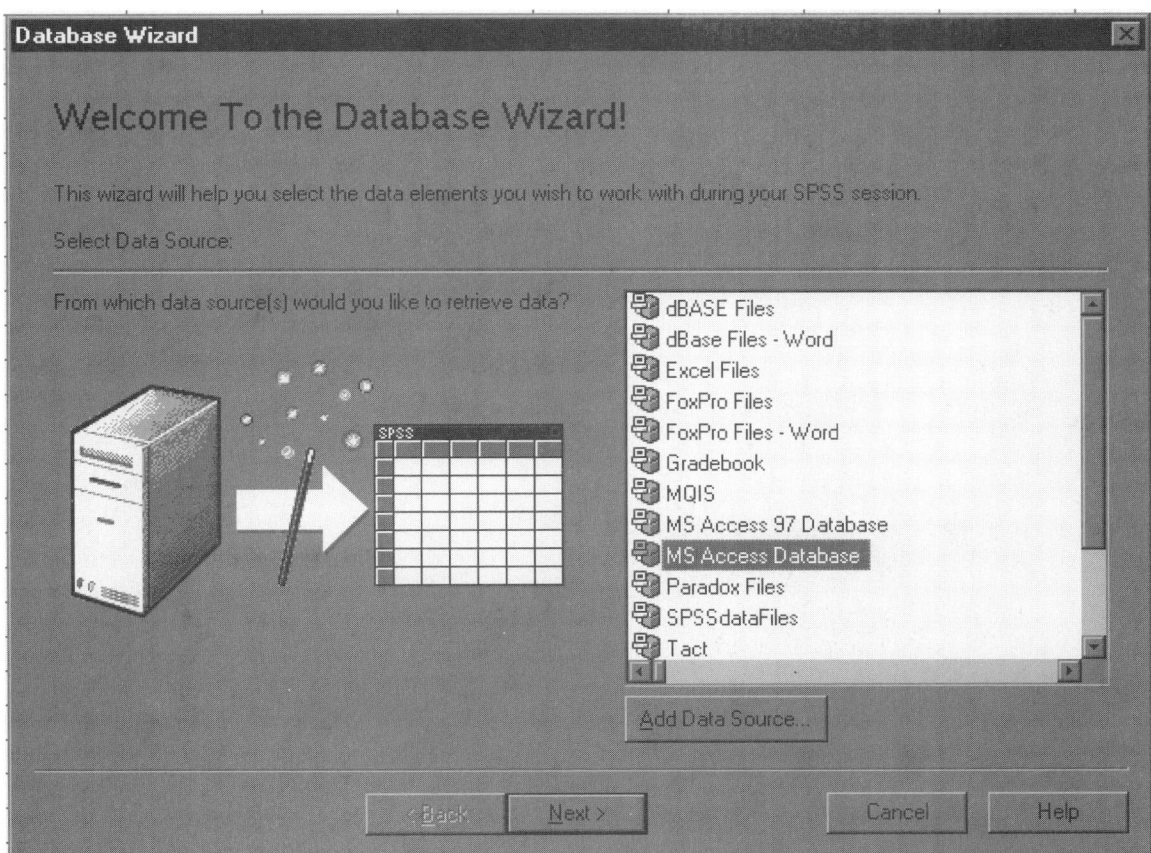

The SPSS Database Wizard opens.

The first time you see the Database Wizard's opening screen, your new database will not be listed—you will add it as a data source in the next section. After the database is added, it will appear in the list of possible data sources.

A-3a Adding a Data Source

In the Database Wizard's opening screen, click Add Data Source. The ODBC Data Source Administrator dialog box should appear. Click on the System DSN tab, and the System DSN page of the dialog box opens. It will look similar to the one in this figure:

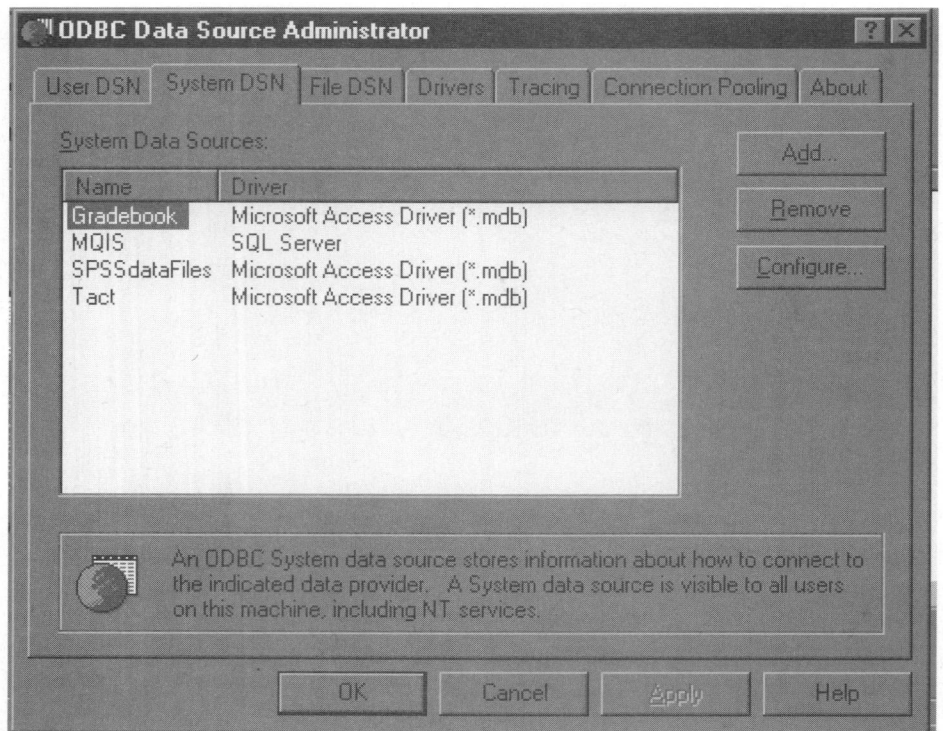

Adding a data source through ODBC.

Click the Add… button, and the Create New Data Source dialog box appears. First, select a driver for your data source: since the relational database was created using Microsoft Access, choose Microsoft Access Driver from the list of names, as shown in this figure:

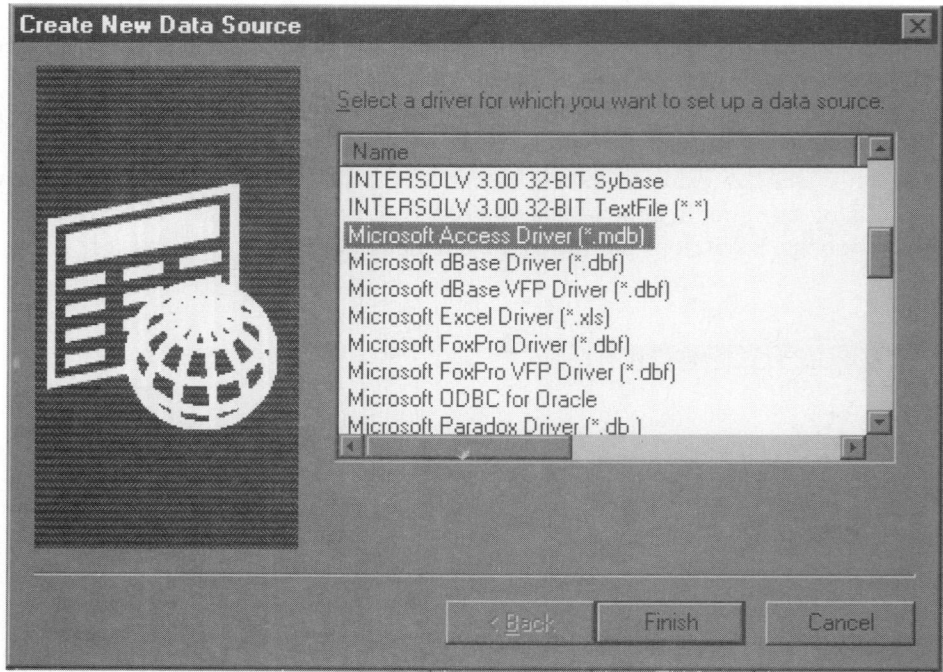

Adding a data source through ODBC, Screen Two.

Click the Finish button, and the ODBC Microsoft Access Setup dialog box appears (see the following figure):

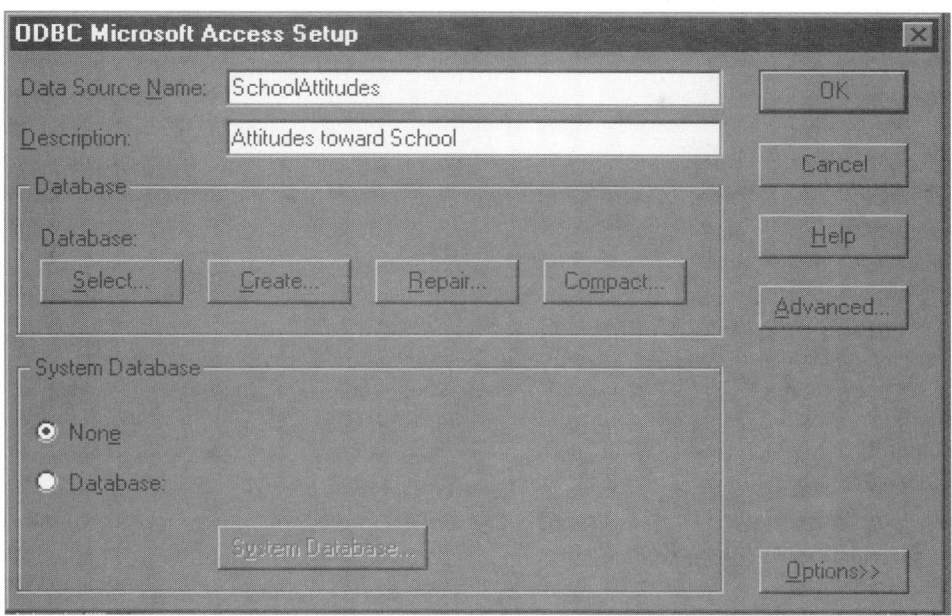

Adding a data source through ODBC, Screen Three.

The name that is entered in the Data Source Name box will appear in the list of data sources in the Database Wizard's opening screen. Our researcher entered SchoolAttitudes in that box, and then entered Attitudes toward School in the Description box. You should do the same. The Description is optional, but it is always a good idea to document thoroughly. The next step is to select the database; click the Select button to proceed.

The Select Database dialog box opens. In the Directories (c:\databases) box, select the folder in which you saved your database. Double-click on the name of the database—Schools.mdb, in this case. The screen should look something like the following figure:

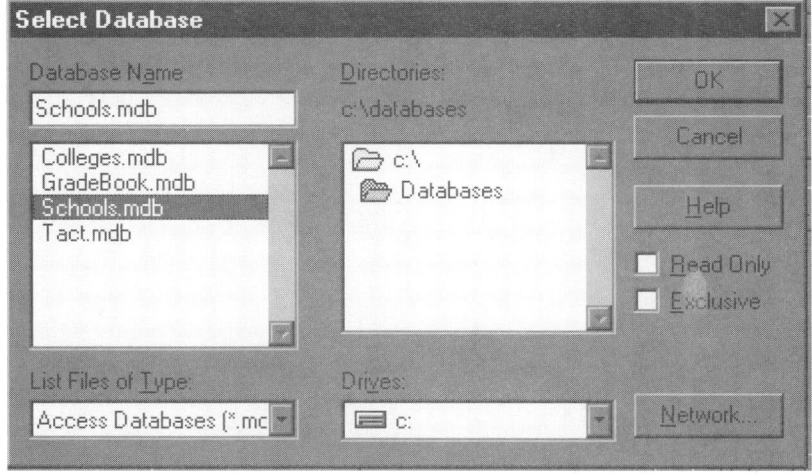

Adding a data source through ODBC, Screen Four.

Click OK. The Database Wizard welcome screen reappears, and your data source (SchoolAttitudes) should now be added to the list, as this figure shows:

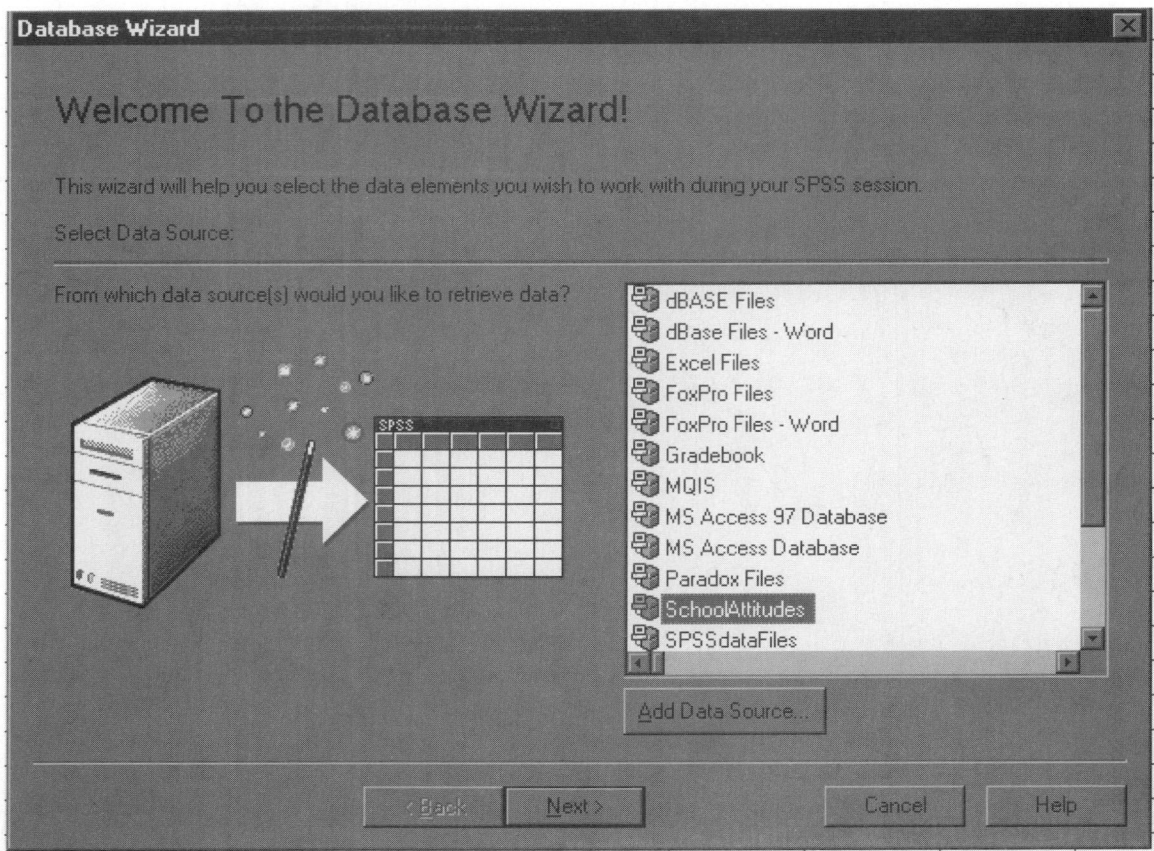

The data source is added to the list.

Remember that these steps need only be done once.

A-3b Selecting Tables from a Data Source

Select SchoolAttitudes from the data source list, and click the Next button.

In Step 2 of the Database Wizard, click and drag all four tables from the Available Tables box on the left to the Retrieve Fields In This Order box on the right. You will use them all in the construction of the data file. The resulting screen should look something like the following figure:

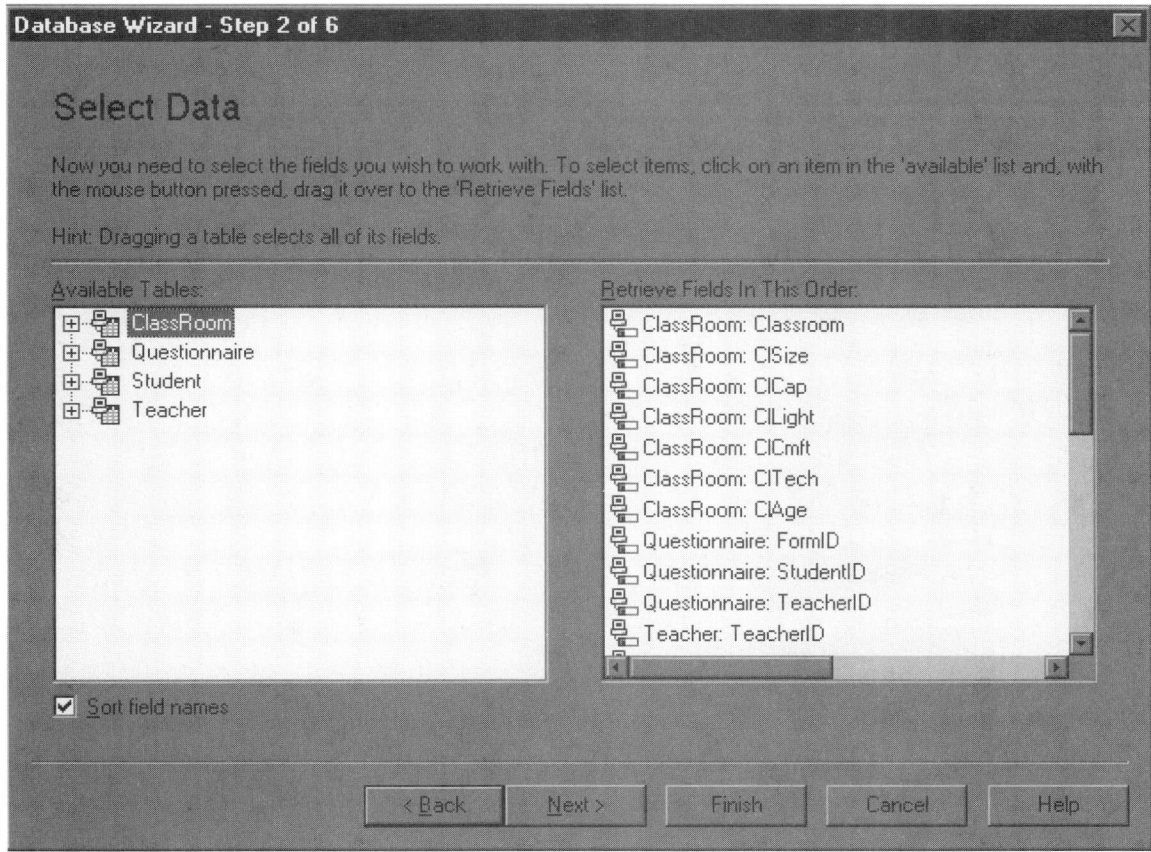

Creating a data file from tables in a data source.

Click Next, and Step 3 of the Database Wizard appears. The relationships between the tables need to be specified. Four boxes corresponding to the four tables in the database should appear (see the following figure). The variables in each table are listed. To specify a relationship, link two tables by dragging one variable name in one table to the corresponding variable name in the other. In this case, each of three variable names in the Questionnaire table is linked to a similar variable name in one of the other three tables:

1. StudentID links to StudentID in the Student table.

2. TeacherID links to TeacherID in the Teacher table.

3. ClassRoomID links to Classroom in the Classroom table.

Note that ClassRoomID is linked with Classroom, demonstrating that the variables do not necessarily have to share the same name.

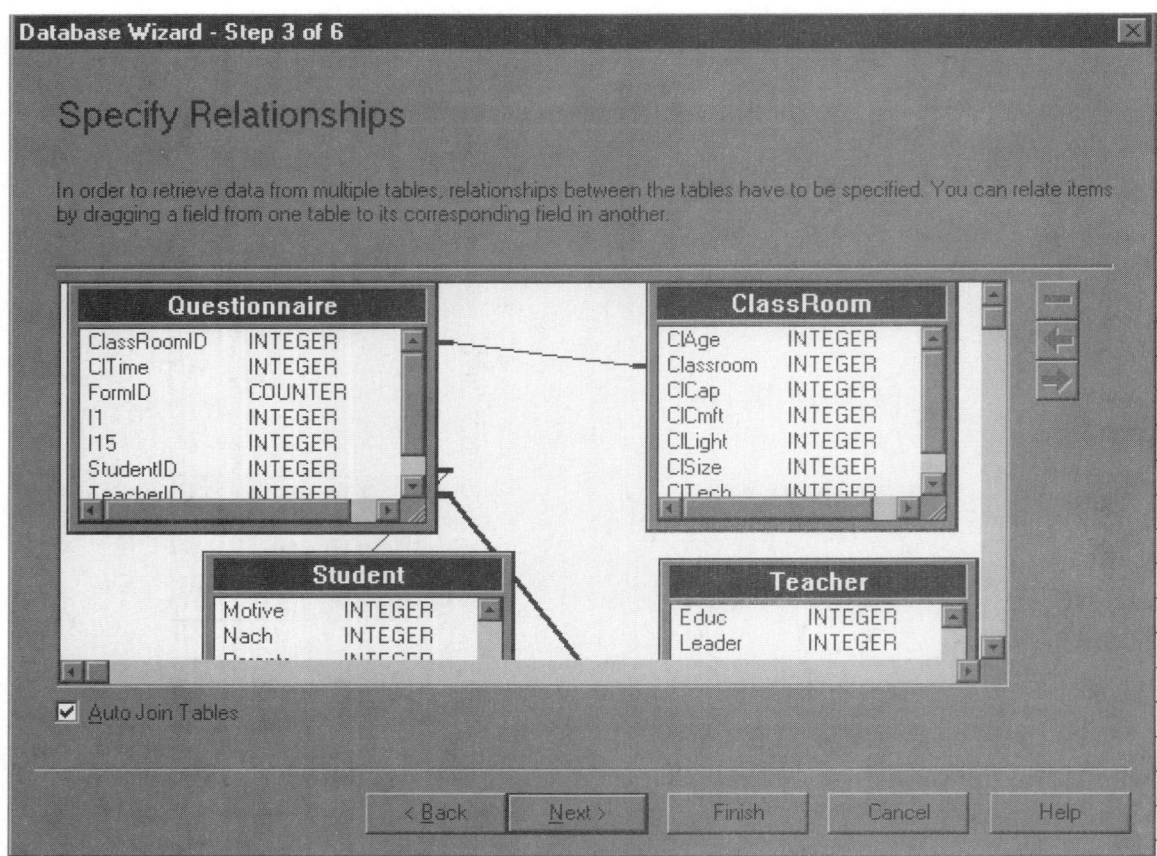

Describing relationships between database tables in SPSS.

Click the Finish button, and the data file magically appears in SPSS, as shown in the following figure:

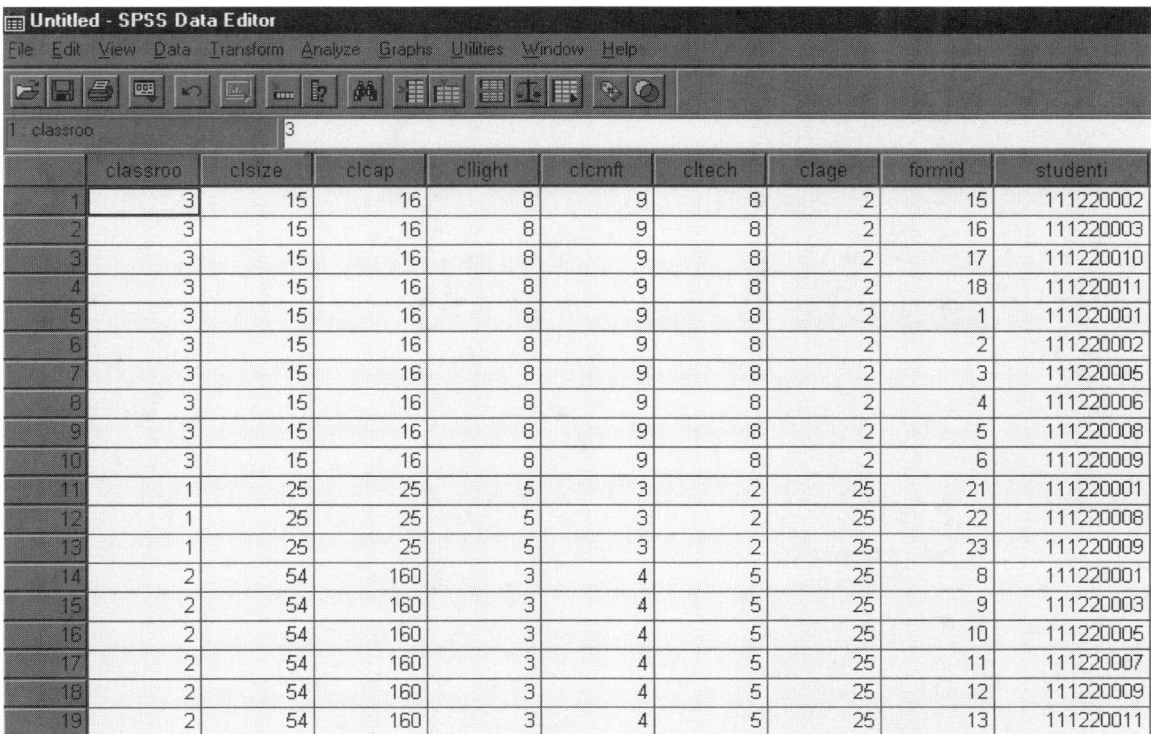

	classroo	clsize	clcap	cllight	clcmft	cltech	clage	formid	studenti
1	3	15	16	8	9	8	2	15	111220002
2	3	15	16	8	9	8	2	16	111220003
3	3	15	16	8	9	8	2	17	111220010
4	3	15	16	8	9	8	2	18	111220011
5	3	15	16	8	9	8	2	1	111220001
6	3	15	16	8	9	8	2	2	111220002
7	3	15	16	8	9	8	2	3	111220005
8	3	15	16	8	9	8	2	4	111220006
9	3	15	16	8	9	8	2	5	111220008
10	3	15	16	8	9	8	2	6	111220009
11	1	25	25	5	3	2	25	21	111220001
12	1	25	25	5	3	2	25	22	111220008
13	1	25	25	5	3	2	25	23	111220009
14	2	54	160	3	4	5	25	8	111220001
15	2	54	160	3	4	5	25	9	111220003
16	2	54	160	3	4	5	25	10	111220005
17	2	54	160	3	4	5	25	11	111220007
18	2	54	160	3	4	5	25	12	111220009
19	2	54	160	3	4	5	25	13	111220011

The data file resulting from the combination of relational database tables.

This data file can be cleaned up by deleting unneeded variables, adding documentation in value and variable labels, and changing variable names. In all respects it can be treated like any other SPSS data file.

Summary

A relational database has many advantages over flat data tables when dealing with complex data. In this chapter you constructed a simple database and then combined it into a single data table in SPSS. The purpose of this exercise was to illustrate the power of using relational databases as a source of data for statistical analysis. You should be aware, however, that the power of these techniques has only been slightly revealed. When large databases from different sources are combined, powerful results are possible.

Appendix

B

References

Bernstin, Peter L. (1996). *Against the Gods*. New York: John Wiley & Sons, Inc.

Brandon, Paul (2000). *Intro Behavior Analysis Laboratory Manual* part 2: Animal Experiments. URL http://www.Mankato.msus.edu/dept/psych/LM2.html . Accessed 10/07/2000

Harvey, John T. (1998). Heuristic Judgment Theory. *Journal of Economic Issues* 32(1), 47–64.

Kain, Richard Y. (1972). *Automata Theory: Machines and Languages*. New York: McGraw-Hill.

Kyburg, Henry E. and Howard E. Smokler (1964). *Studies in Subjective Probability*. New York: John Wiley & Sons, Inc.

Letcher, John S. Jr., John K. Marshall, James C. III Oliver, and Nils Salvesen (1987). Stars & Stripes. *Scientific American* 257(2), 34–40.

Michell, J. (1986). Measurement Scales and Statistics: A Clash of Paradigms. *Psychological Bulletin* 100–3, 398–407.

Paulos, John Allen (1988). *Innumeracy: Mathematical Illiteracy and Its Consequences*. New York: Hill and Wang.

Peterson, Ivars (1998). *The Jungles of Randomness: A Mathematical Safari*. New York: John Wiley & Sons, Inc.

Stevens, S. S. (1951). *Mathematics, Measurement and Psychophysics*. In S. S. Stevens (Ed.), *Handbook of Experimental Psychology* (pp. 1–49). New York: John Wiley & Sons.

Stockburger, David W. (2000). *Introductory Statistics: Concepts, Models, and Applications*. Cincinnati, Ohio: Atomic Dog Publishing.

Tuft, Edward R. (1983). *The Visual Display of Quantitative Information*. Cheshire, Connecticut: Graphics Press.

Tversky, Amos and Daniel Kahneman (1974). Judgment under Uncertainty: Heuristics and Biases. *Science* 185, 1124–1131.

Wenman, Anders, Peter Juslin, and Mats Bjorkman (1998). The Confidence-Hindsight Mirror Effect in Judgment: An Accuracy-Assessment Model for the Knew-It-All-Along Phenomenon. *Journal of Experimental Psychology: Learning, Memory, and Cognition* 24(2), 415–431.

Glossary

Symbols

μ **(mu)**   One of two parameters of normal curves.

σ **(sigma)**   One of two parameters of normal curves.

Σ **(summation sign)**   Used to represent summation in an expression.

A

absolute cumulative frequency   The number of scores that fall at or below a given score value. It is computed by adding up the number of scores that are equal to or less than a given score value.

absolute frequency   The actual number of times that a particular score (or score with an interval) appears in the data.

absolute frequency polygon   Frequency distribution that is drawn exactly like a histogram except that points are drawn rather than bars. The points are plotted on the graph at the intersection of the midpoint of the interval and the height of the frequency. When the points are plotted, the dots are connected with lines, resulting in a frequency polygon.

absolute measure   A measure of some quantity, such as weight.

absolute value   The positive value of any score.

additive component   The a in the linear transformation equation $X' = a + bX$. The constant that is added in a linear transformation.

algebra   A formal symbolic language, composed of strings of symbols. The symbol set of algebra includes numbers, variables, operators, and delimiters. In combination, they define all possible sentences that may be created in the language.

alpha   The probability of rejecting the null hypothesis when in fact the null hypothesis is true. The probability of deciding that the effects are real when in fact the results were due to chance. Alpha is directly set by the researcher.

analysis of variance   A hypothesis testing procedure that tests for effects by comparing two or more means.

ANOVA   *See* analysis of variance.

apparent limits   The original beginning and ending values of an interval. For example, for the interval 33 to 35, the value of 33 would be called the apparent lower limit, and the value of 35 would be the apparent upper limit.

area under a curve   The total area underneath a curve defined using a mathematical equation. It represents theoretical relative frequency or probability.

area under theoretical models of distributions   A method of estimating probabilities.

B

bar graph   Similar to a histogram, but the data are nominal categorical in nature, and the bars do not touch.

Bayesian Statistics   A branch of statistics whose foundation is the recomputation of probabilities based on data.

bimodal   A distribution with two different scores occurring the same number of times with the greatest frequency.

binomial expansion   A special form of exponential notation that occurs when a phrase connected with addition or subtraction operators is raised to the second power.

bivariate data   Data that contains two variables (x, y). Also called paired data.

C

causal variable   Changes in values of this variable are directly related to changes in the variable it causes.

causation The establishment of a direct link between two variables, usually done using the experimental method.

Central Limit Theorem Relates the sampling distribution of the mean to the theoretical model of the distribution of scores. The Central Limit Theorem comes in a variety of flavors, but generally stated says that the sampling distribution of the mean will be a normal distribution with a theoretical mean equal to mu and a theoretical standard deviation, called the standard error, equal to sigma of the model of scores divided by the square root of the sample size. In theory the Central Limit Theorem requires that the sample size approach infinity, but in practice the results converge with relatively small sample sizes ($N > 10$).

central tendency A typical or representative score. Mean, median, and mode are measures of central tendency.

chi-square statistic A measure of the difference between observed and expected values.

chi-squared distribution A theoretical probability model, described by a single parameter, called degrees of freedom. In this model, scores are positively skewed and range from zero to infinity.

compound event A combination of simple events joined with either "and" or "or."

compound probabilities The probability of a compound event.

computational formula for the standard error of estimate A formula to compute the standard error of estimate that includes the variance of Y and the correlation coefficient. It is easier to compute than the definitional formula because it does not require a table of squared residuals to be computed.

conditional distribution A distribution of a variable given a particular value of another variable.

conditional probability The probability of an event given that another event is true.

confidence interval A pair of scores that describe a theoretical range of values of a score.

constant A value that does not change with the different values for the counter variable (i).

contingency tables Frequency tables of two variables presented simultaneously. Constructed by listing all the levels of one variable as rows in a table and the levels of the other variables as columns.

control condition In an experiment, a condition identical to the treatment condition except no treatment is given.

correlation Changes in one variable are related to changes in another variable—they "co-relate."

correlation coefficients Numbers between minus one and one that measure the linear relationship between two variables.

correlation matrix A table of all possible correlation coefficients between a set of variables.

crossed design Experimental design in which each subject sees each level of the treatment condition.

cumulative frequency Found from the absolute frequency by either adding up the absolute frequencies of all scores smaller than or equal to the score of interest, or by adding the absolute frequency of a score value to the cumulative frequency of the score value immediately below it.

cumulative frequency polygon Plot scores on the X-axis, and the absolute cumulative frequency on the Y-axis. The points are plotted at the intersection of the upper real limit of the interval and the absolute cumulative frequency.

D

data ink A large number of lines or bars in a histogram that make the comprehension of the graph difficult.

degrees of freedom (df) The number of scores that are free to vary.

delimiters Punctuation marks in algebra.

descriptive function of statistics Procedures for organizing and describing sets of data.

distorted sample A sample which, by chance, fails to reflect the entire population from which it was drawn.

distribution A set of numbers collected from a well-defined universe of possible measurements arising from a property or relationship under study.

E

effect When a change in one thing is associated with a change in another. The changes may be either quantitative or qualitative.

estimator A statistic used to estimate a model parameter.

exact significance level The probability of finding an effect equal to or larger than the effect found in the study given that the null hypothesis is true.

expected utility theory A mathematical theory combining cost and probabilities.

experimental condition In an experiment, the level of treatment in which some treatment is given.

experimental design The manner in which an experiment is set up. Specifically, the way the treatments are administered to subjects.

exponential notation An example of a shorthand notational scheme, one of a number of rewriting rules that exist within algebra to simplify an algebraic phrase with a shorthand notation of that phrase.

F

F-distribution A theoretical probability distribution characterized by two parameters, df_1 and df_2, both of which affect the shape of the distribution. The distribution is nonsymmetrical, skewed in the positive direction.

formal language Languages such as mathematics, logic, and computer languages. Statistics is a branch of the formal language of mathematics, algebra.

form-board test One of the earliest psychological tests where the score of the person being tested is the time it takes to place a number of pegs in a board of cutout forms.

fractions An algebraic phrase involving two numbers connected by the operator /. For example, 7/8.

F-ratio The Mean Squares Between divided by the Mean Squares Within. A measure of how different the means are relative to the variability within each sample.

frequency distributions Pictures of data. One method of describing data.

frequency polygon The actual drawing of a frequency distribution.

frequency table The first step in drawing a frequency distribution. A way of organizing the data by listing every possible score (including those not actually obtained in the sample) as one column of numbers and the frequency of occurrence of each score as another column of numbers.

G

grouped frequency polygons Generally used when a table and graph present so much data that information can't be easily discerned. Use intervals to group data, with some loss of information (precision) about the data in order to gain understanding about the distribution.

H

histogram A graph drawn by plotting the scores (midpoints) on the X-axis and the frequencies on the Y-axis, and drawing a bar for each score value, the width of the bar corresponding to the real limits of the interval and the height corresponding to the frequency of the occurrence of the score value.

hypothesis tests Procedures for making rational decisions about the reality of effects.

I

inferential function of statistics Procedures by which you take a sample from the population, describe the numbers of the sample using descriptive statistics, and infer the population distribution.

intercept Another name for the additive component in a linear transformation. When a line is drawn on a plane, the line crosses the Y-axis at the intercept. The a value that defines where the line crosses the Y-axis in a regression model.

interval estimate *See* confidence interval.

interval scales Measurement systems that possess the properties of magnitude and intervals, but not the property of rational zero.

intervals (property of) A property of the measurement system that is concerned with the relationship of differences between objects. If a measurement system possesses the property of intervals, the unit of measurement means the same thing throughout the scale of numbers.

invariant Does not change.

inverse relationship A relationship between two variables where, in general, as one variable becomes larger, the other becomes smaller.

IQ scale Test scores have a mean of 100 and a standard deviation of either 15 or 16, depending upon the test selected.

L

least-squares criterion   A value that minimizes the sum of squared differences between the scores and the predicted values.

linear regression   A prediction model of the form $Y' = a + bX$.

linear transformations   Transformations where each score is multiplied by a constant and then a different constant is added to the resulting product.

M

magnitude (property of)   A property of the measurement system whereby an object that has more of the attribute than another object is given a bigger number by the rule system.

marginal frequencies   In a contingency table, the values of the cell frequencies summed across both the rows and the columns and placed in the margins.

mean   The sum of the scores divided by the number of scores. The most often used measure of central tendency.

Mean Squares Between (MS_B)   The variance of the means times the number of scores within each group, an estimate of the theoretical variance of scores.

Mean Squares Within (MS_W)   The mean of the variances, an estimate of the theoretical variance of scores.

measurement   In mathematical terms, a functional mapping from a set of objects into the set of real numbers. The act of applying the rules of a measurement system. A process in which the symbols of the language of algebra are given meaning.

measurement system   Any set of rules for assigning numbers to attributes of objects.

median   The score value that cuts the distribution in half, such that half the scores fall above the median and half fall below it. A measure of central tendency.

mode   The most frequently occurring score value. On a frequency distribution, it is the score value that corresponds to the highest point. A measure of central tendency.

model   A representation containing the essential structure of some object or event in the real world. There are two types—physical and symbolic.

model of sample frequency distribution   *See* probability model.

mu (μ)   One of two parameters of normal curves. It defines the center of the distribution.

multiple t-tests   Hypothesis testing procedure when there are more than two groups that compares all possible pairs of means using a t-test.

multiplicative component   The b in the linear transformation equation $X' = a + bX$. The constant that is multiplied times the score in a linear transformation.

N

natural language   Languages such as English, German, French, and Spanish.

negative correlation coefficient   If one variable increases, the other variable decreases, and if one decreases, the other increases.

negative exponential distribution   A probability model often used to model real-world events, which are relatively rare, such as the occurrence of earthquakes or winning a lottery.

negatively skewed distribution   An asymmetrical distribution that points in the negative direction, with the mean being smaller than the median, which is smaller than the mode.

nested design   Experimental design in which each subject receives one, and only one, treatment condition.

nested t-test   A hypothesis testing procedure for nested designs with two levels.

nominal-categorical scale   Nominal scale in which objects are grouped into subgroups and each object within a subgroup is given the same number. The subgroups must be mutually exclusive—that is, an object may not belong to more than one category or subgroup.

nominal-renaming scale   Nominal scale in which each object in the set is assigned a different number, that is, renamed with a number.

nominal scales   Measurement systems that do not possess the properties of magnitude, intervals, and rational zero. Nominal scales may be subdivided into two groups: nominal-renaming and nominal-categorical.

noncumulative polygons   Absolute or absolute cumulative frequency polygons.

nonoptimal regression model   A regression model that does not meet the least squares criterion.

normal curve *See* normal distribution.

normal distribution A probability model commonly called the normal curve. It is symmetric with scores more concentrated in the middle than in the tails (ends).

null hypothesis The hypothesis that there are no effects. Chance or random variation is responsible for any differences discovered.

O

one-tailed t-test A direction t-test where alpha is placed in a single tail of the distribution under the null hypothesis.

operators In the language of algebra, symbols that signify relationships between numbers and/or variables.

optimal regression model A regression model that meets the least squares criterion.

ordinal scales Measurement systems that possess the property of magnitude, but not the property of intervals.

outlier A score that falls outside the range of the rest of the scores on the scatterplot.

overlapping cumulative frequency polygons The polygons for each group are drawn on the same set of axes, distinguished with different types of lines.

overlapping frequency distributions Two frequency distributions that share the same X-axis, allowing direct comparison of their shapes.

overlapping relative cumulative frequency polygons Two cumulative frequency distributions that share the same X-axis, allowing direct comparison of their shapes.

overlapping relative frequency polygons The polygons for each group are drawn on the same set of axes, distinguished with different types of lines.

P

parameters Variables within the model that must be set before the model is completely specified. Variables that change the shape of the probability model.

parsimonious models Simple models that have a great deal of explanatory power.

path analysis A branch of correlational analysis that attempts to establish causation from correlational evidence.

percentile rank based on the normal curve The percentage of scores that fall below a given score in a hypothetical distribution of scores based on the normal curve.

percentile rank based on the sample The percentage of scores that fall below a given score within a sample of scores.

percentile ranks The percentage of scores that fall below a given score.

point estimate A single value that represents the best predicted value of Y.

population *See* probability model.

population distribution A theoretical probability model.

positive correlation coefficient If one variable increases (or decreases), the other variable also increases (or decreases).

positively skewed distribution An asymmetrical distribution that points in the positive direction, with the mode smaller than the median, which is smaller than the mean.

predicted variable The variable being predicted. The dependent variable.

predictor variable The variable used to predict. The independent variable.

probability The theoretical relative frequency of the event in a model of the population. The area under the curve between any two points.

probability density function (pdf) *See* probability model.

probability distribution *See* probability model.

probability model A mathematical equation used to model a relative frequency distribution. Describes the likelihood of a specific event associated with a specific process. Attempts to capture the essential structure of the real world by asking what the world might look like if an infinite number of scores were obtained and each score was measured infinitely precisely. Also called theoretical probability distribution, probability density function (pdf), population, model of the sample frequency distribution, and probability distribution.

probability theory A mathematical model of uncertainty. Defines probabilities of simple events in algebraic terms and then presents rules for combining the probabilities of simple events into probabilities of complex events given that certain conditions are present (assumptions are met).

probable error   Inferential statistics allows you to specify an amount of error that may have skewed the results of a sample.

property of intervals   A property of the measurement system that is concerned with the relationship of differences between objects. If a measurement system possesses the property of intervals, the unit of measurement means the same thing throughout the scale of numbers.

property of magnitude   A property of the measurement system whereby an object that has more of the attribute than another object is given a bigger number by the rule system.

property of rational zero   A property of the measurement system whereby an object that has none of the attributes in question is assigned the number zero by the system of rules. The object does not need to really exist. This property is necessary for ratios between numbers to be meaningful, but is not necessary to make meaningful inferences in many application of statistics.

R

random sample   A sample in which each individual in the population was likely to be included in the sample.

range   A measure of variability. The largest score minus the smallest score.

rational zero (property of)   A property of the measurement system whereby an object that has none of the attributes in question is assigned the number zero by the system of rules. The object does not need to really exist. This property is necessary for ratios between numbers to be meaningful, but is not necessary to make meaningful inferences in many application of statistics.

ratio scales   Measurement systems that possess all three properties: magnitude, intervals, and rational zero.

raw score   The score that is given.

real limits   The midpoints between midpoints of intervals. Two points that function as cut-off points for a value, or the midpoints between the values. (A value would be the midpoint of the interval.) The interval 36–38 would therefore have a real lower limit of 35.5 and a real upper limit of 38.5. Each interval has a real lower limit and a real upper limit. The difference between the real limits of an interval is equal to the interval size.

regression   A movement backward toward the mean.

regression analysis   Application of linear regression procedures, including parameter and error estimation techniques.

regression coefficients   The values of the regression weights.

regression line   The representation of the regression model on a scatterplot.

regression model   Used to predict one variable from one or more other variables.

relational database   A number of flat tables linked together with index variables. Complex queries and tables can be constructed with relational databases.

relative cumulative frequency   The proportion of scores that fall at or below a given score value. It is computed by dividing the absolute cumulative frequency by the number of scores (N).

relative cumulative frequency polygon   Plot scores on the X-axis and the relative cumulative frequency on the Y-axis. Points are plotted at the intersection of the upper real limit and the relative cumulative frequency.

relative frequency   The proportion of scores that have a particular value. Computed by dividing the frequency of a score by the number of scores (N).

relative frequency polygon   Drawn exactly like the absolute frequency polygon except the Y-axis is labeled and incremented with relative frequency rather than absolute frequency.

relative measure   A measure of a variable relative to some other measure. The ratio of weight to height would be a relative measure of weight.

residuals   Deviations of observed and predicted values.

S

sample   A representative set of individuals that shows the quality or nature of the whole from which it was taken.

sample distribution   The distribution resulting from the collection of actual data.

sample statistics   Mathematical equation used to measure properties of samples. Sample statistics are used as estimators of parameters in the probability models.

sampling   The procedure for acquiring a sample.

sampling distribution   A theoretical distribution of a sample statistic.

scale types   Originally proposed as a way to classify measurement systems with respect to whether the properties would be preserved when various mathematical operations were used with the numbers that the system produced. Also used to classify measurement systems with respect to appropriateness for various kinds of statistical analysis. Four types are discussed in this text: nominal, ordinal, interval, and ratio.

scatterplot   A visual representation of the relationship between the X and Y variables.

scientific method   A way of creating, verifying, and modifying models to simplify and explain the complexity of the world. A procedure for the construction and verification of models.

sigma (σ)   One of two parameters of normal curves. It defines the spread or dispersion of the distribution.

significance level   *See* alpha.

skewed distribution   A distribution that is asymmetrical, and in which the mean, median, and mode do not all fall at the same point.

slope   Another name for the multiplicative component in a linear transformation. When a line is drawn on a plane, the steepness of the line is determined by the slope. The value of b in the regression equation $Y' = a + bX$.

squared correlation coefficient   The proportion of variance in Y that can be accounted for by knowing X.

standard deviation   A measure of variability. The positive square root of the variance.

standard error   The theoretical standard deviation of a sampling distribution.

standard error of estimate   A measure of error in prediction.

standard normal curve   A member of the family of normal curves with $\mu = 0.0$ and $\sigma = 1.0$.

standard scores   A linear transformation such that the transformed mean and standard deviation are 0 and 1 respectively. Also called z-scores.

stanine transformation   Scores are linearly transformed to a distribution with a mean of 5 and a standard deviation of 2 and the decimals are dropped, so that the numbers are integers between 0 and 9.

statistics   Summary numbers used to describe other numbers. One method of describing data.

statistics (descriptive function of)   Procedures for organizing and describing sets of data.

statistics (inferential function of)   Procedures by which you take a sample from the population, describe the numbers of the sample using descriptive statistics, and infer the population distribution.

subjective probabilities   Probabilities obtained by procedures designed to extract "degree of belief" from individuals.

subsamples   Partitions of a measured variable. A dichotomous variable would be partitioned into two subsamples, one for each level of the variable.

subscripted variables   A method by which large numbers of variables can easily be represented. Its form is X_i, where the X is the variable name and the subscript (i) is a counter variable that can take on values from 1 to N.

summation notation   A scheme that provides a means of representing both a large number of variables and the summation of an algebraic expression.

summation sign (Σ)   Used to represent summation in an expression.

sum of squared deviations   The sum of the squared differences between the observed and predicted values of Y.

symbolic model   A model constructed using either a natural or formal language.

symmetrical distribution   A distribution in which the mean, median, and mode all fall at the same point. If drawn, cut out, and folded, the two sides would be identical.

syntax of language   The set of rules that determines which strings of symbols form sentences and which do not language.

T

t distribution   A theoretical distribution that is symmetrical, bell-shaped, has tails approaching the X-axis but never touching, and total area under the curve equal to one. The t distribution has three parameters, degrees of freedom, mu, and sigma. The fewer the degrees of freedom, the flatter the t distribution is relative to the normal distribution.

theoretical probability distribution   *See* probability model.

transformation   A procedure that converts a number into another number.

transformational rules Rules for rewriting sentences in the language of algebra without changing their meaning, or truth value.

transformed scores Raw scores that have been converted into another number. Generally transformed scores can be more easily interpreted than raw scores.

treatment Quantitatively or qualitatively different levels of experience.

treatment condition Any of the levels of treatment in an experiment.

T score Score that has been transformed into a scale with a mean of 50 and a standard deviation of 10.

t-test A hypothesis test employing the t distribution.

two-tailed t-test Alpha is divided in half and placed in both tails of the distribution under the null hypothesis.

Type I error The null hypothesis is rejected when in fact it is true. The hypothesis testing procedure decides that the effects are real when if fact the results were due to chance.

Type II error The null hypothesis is retained when in fact the alternative hypothesis is true. The hypothesis testing procedure decides that the no effects model could explain the results when in fact the effects were real.

U

unbiased estimator A statistic whose theoretical mean of the sampling distribution is equal to the value of the parameter it is estimating.

uniform distribution A probability model that is shaped like a rectangle, where each score is equally likely. Useful model when the phenomenon being modeled is relatively stable over a range of values.

utility The gain or loss experienced by a player depending upon the outcome of the game.

V

variability The spread or dispersion of scores. Three measures of variability are the range, the variance, and the standard deviation.

variables In the language of algebra, symbols that stand for any number.

variance A measure of variability. A measure of score dispersion.

vectors Lines from the origin to a point on a graph, sometimes represented as points on a graph.

Z

z-scores *See* standard score.

Index

review of algebra, 14
rewriting rules, 17, 19, 90
rounding error, 46, 105, 162
row marginals, 50
ruler, 27
rules of precedence, 15

S

sample, 3, 25, 30, 40, 64, 67, 94, 96, 98, 103, 162, 169–170, 176, 182, 186–189, 192, 194–195, 254
sampling, 3, 176, 186–189, 192–195, 254–256
sampling distribution, 176, 186–189, 192–195, 254–256
scale, 11–12, 23–28, 46, 56, 162, 169, 254
scale types, 24, 26–28
scatterplot, 158
scientific method, 6, 8, 11–12
scores, 112
selecting the interval size, 56
sentence, 7–8, 14–16, 19–20
sentences in algebra, 15
shoe size, 30–34, 37, 46, 49, 64–65, 67, 107–110, 162–163, 179, 182
shorthand notation, 17, 22, 86–87
short-term memory, 2
sign, 85–90
sign of the correlation coefficient, 158, 168
significance, 195
significant, 175, 182, 256
simplification, 8, 12, 20, 88, 90
simplifying assumptions, 9
simulation, 11

size, 55–56, 58–59, 61–62, 170, 189, 194–195
skewed, 99, 100–101
skewness, 98
slope, 41, 161–162, 171
sophistication, 14
sort, 36
special education classroom, 27
SPSS data file, 35
SPSS, 31–43, 47, 49, 54, 57–58, 105–110, 166–168
square root of N, 189
squared, 163, 164, 171
standard deviation, 25, 102–105, 107–110, 162, 166, 186–189
standard error, 164, 188
standardized intelligence test, 27, 192
statistical calculator, 104, 158, 164
statistician, 22, 28, 65, 174, 187–188, 192, 194
statistics, 167, 169
statistics, 2–3, 6, 8–9, 24–27, 32, 71, 86, 94, 102, 105–106, 110, 175–176, 179, 182, 186–188
straight line, 158
strings of symbols, 7, 14, 20
subsamples, 46, 51
subscript, 22, 86, 188
subscripted variables, 86, 89
sum of the product, 88, 179
summary numbers, 2, 30
summation notation, 90
symbol set, 7, 14, 20
symbolic model, 7, 12
symmetric, 70, 98–99, 166
symmetrical distribution, 98–99
syntax, 7, 14–15

T

table of means and standard deviations, 168
theoretical mathematician, 22
thought experiment, 176, 187–188, 192, 254
three dots, 86
total variance, 163, 164
transform, 162
transform, 7, 162
transformation, 8, 12, 16, 162

U

umbrella, 175, 182
unbiased estimator, 188
uniform distribution, 65, 187
unit of measurement, 23, 56, 58, 104
units of measurement squared, 104
upper limit, 56, 58
upper real limit, 32, 41, 42

V

variability, 94, 102–105, 109–110
variable name, 32, 86, 91, 166
variables, 14, 86, 102–105, 110
variance, 103, 162–164, 188
verification, 8, 12

W

wing, 10
winged keel, 12
word problems, 8

Z

z-score transformation, 162
z-score, 161–162, 165